AI FOR DIVERSITY

Artificial intelligence (AI) is increasingly impacting many aspects of people's lives across the globe, from relatively mundane technology to more advanced digital systems that can make their own decisions. While AI has great potential, it also holds great peril depending on how it is designed and used. *AI for Diversity* questions how AI technology can lead to inclusion or exclusion for diverse groups in society. The way data is selected, trained, used, and embedded into societies can have unfortunate consequences unless we critically investigate the dangers of systems left unchecked, and can lead to misogynistic, homophobic, racist, ageist, transphobic, or ableist outcomes. This book encourages the reader to take a step back to see how AI is impacting diverse groups of people and how diversity-awareness strategies can impact AI.

Roger A. Søraa is an Associate Professor at the Department of Interdisciplinary Studies of Culture at the Norwegian University of Science and Technology (NTNU), in Trondheim, Norway. His main research interests are the digitalization and robotization of society and its ethical, gendered, and epistemological consequences. He leads the Digitalization and Robotization of Society research group and coordinates several research projects, including the Horizon Europe project "BIAS: Mitigating Diversity Biases in the Labor Market."

AI FOR EVERYTHING

Artificial intelligence (AI) is all around us. From driverless cars to game-winning computers to fraud protection, AI is already involved in many aspects of life, and its impact will only continue to grow in the future. Many of the world's most valuable companies are investing heavily in AI research and development, and not a day goes by without news of cutting-edge breakthroughs in AI and robotics.

The AI for Everything series will explore the role of AI in contemporary life, from cars and aircraft to medicine, education, fashion, and beyond. Concise and accessible, each book is written by an expert in the field and will bring the study and reality of AI to a broad readership including interested professionals, students, researchers, and lay readers.

AI for Cars
Josep Aulinas & Hanky Sjafrie

AI for Digital Warfare
Niklas Hageback & Daniel Hedblom

AI for Art
Niklas Hageback & Daniel Hedblom

AI for Creativity
Niklas Hageback

AI for Death and Dying
Maggi Savin-Baden

AI for Radiology
Oge Marques

AI for Games
Ian Millington

AI for School Teachers
Rose Luckin, Karine George &
Mutlu Cukurova

AI for Learning
Carmel Kent & Benedict du Boulay

AI for Social Justice
Alan Dix & Clara Crivellaro

**AI for the Sustainable
Development Goals**
Henrik Skaug Sætra

AI for Healthcare Robotics
Eduard Fosch-Villaronga & Hadassah
Drukarch

AI for Physics
Volker Knecht

AI for Diversity
Roger A. Søraa

For more information about this series please visit:
https://www.routledge.com/AI-for-Everything/book-series/AIFE

AI FOR
DIVERSITY

ROGER A. SØRAA
Norwegian University of Science
and Technology

CRC Press
Taylor & Francis Group
Boca Raton   New York   London

CRC Press is an imprint of the
Taylor & Francis Group, an **informa** business

First edition published 2023
by CRC Press
6000 Broken Sound Parkway NW, Suite 300, Boca Raton, FL 33487-2742

and by CRC Press
4 Park Square, Milton Park, Abingdon, Oxon, OX14 4RN

CRC Press is an imprint of Taylor & Francis Group, LLC

ISBN: 978-1-032-07444-3 (hbk)
ISBN: 978-1-032-07356-9 (pbk)
ISBN: 978-1-003-20695-8 (ebk)

DOI: 10.1201/9781003206958

Typeset in Joanna
by SPi Technologies India Pvt Ltd (Straive)

CONTENTS

PREFACE

Digital technology has a major impact on our lives. They can be a wonderful addition to enrich, simplify, and optimize daily practices if used well, but they also have more sinister implications when not designed or used responsibly for a diverse group of people. Unwrapping how digital technology can be used to better—not worsen—human lives, and how it can be inclusive to a wide range of users and developers is the centerpiece of this book. The idea for writing this book came from my time working for the European Commission as an expert advisor and evaluator on the *Gendered Innovations 2 Expert Group* on how gender and diversity awareness can improve technologies and research in more just, inclusive, and robust ways. When people from homogeneous backgrounds are left alone to define which programs, systems, and data we use in our daily lives and societies, we risk building a less inclusive and more heavily trajected world toward their priorities and experiences—leaving all "others" behind. We still have a long way to go to diversify AI and its related fields to be more inclusive and better for all. Use this book as a guiding introduction to how you can identify unfair data and situated exclusion and a tool for avoiding bias and building knowledge for a better world for all.

WHO IS THIS BOOK NOT FOR?

If you are looking for a technical book explaining how different AI systems work, this is not the book for you. Neither is this book an encyclopedia of all existing AIs that exist. There are bound to be blind spots and systems not covered in such a short book. Likewise, gender scholars might want a much larger book that goes in-depth on each topic on a more theoretical level. Neither the dystopian reader seeking a vendetta against all AI as bad nor the blissful techno-utopists seeing all AI as benevolent should pick up this book. However, if you want a critical investigation into how different AI systems can impact societal and individual diversity in terms of inclusion and exclusion of specific groups, and want to learn more about that topic, then, by all means, read this book, and help build a better world of AI for diversity.

ACKNOWLEDGMENTS

I would like to thank my research group and research partners who have been involved in the commenting and writing process of this book. Particularly, I would like to thank the helpful reviewers for their insightful comments and suggestions: Ashley Shew, Elisabeth Stubberud, Inga Strümke, Jennifer Branlat, Julie Katrine Flikke, Katherine Evans, Kristine Ask, Mark W. Kharas, Martine Sletten, Mascha Kurpicz-Briki, Siri Øyslebø Sørensen, Sofia Moratti, Sophia Ivarsson, and two anonymous reviewers. Big thanks to Nienke Bruijning for the illustrations and my Editor, Elliott Morsia, and the editorial team at CRC Press/Taylor & Francis Group, as well as project manager Yassar Arafat. This book was made possible by the following projects: Robotics4EU (European Union Horizon 2020 research and innovation program grant agreement number 101017283), LIFEBOTS Exchange (European Union Horizon 2020 research and innova-tion program grant agreement number 824047), and AUTOWORK (Norwegian Research Council project number 301088).

ABBREVIATION LIST

AI	Artificial intelligence
LGBTQ+	Lesbian, gay, bisexual, transgender, queer, plus others
ML	Machine learning
BIPoC	Black, Indigenous, and People of Color
STS	Science and technology studies

1

OPENING THE BLACK BOX
OF AI

Artificial intelligence (AI) is often mystified and misunderstood—especially concerning who benefits and who are excluded when it is used. AI is by techno-utopists seen as benevolent, magnificent techno-transformations that can solve every imaginable problem, for example, health care crises, climate change, cybersecurity, accelerating learning, and protecting biodiversity, to mention some. Techno-dystopians, on the other hand, see it as an eerie harbinger of societal problems that can even lead to humanity's demise, where top-down state or corporate surveillance and control, military doomsday weapons, and biotechnical ethics of designer children run amok. This *technological determinism*—believing that technology determines how society accepts and changes with the uptake and development of new technology, whether it be toward better or worse futures—is damaging the discourse on how we can actually shape and choose how to use AI systems. In this book, I look between these binary understandings of "good or evil AI," and explore more pragmatically what AI entails for us as individual users of it, why developers design it in certain ways, and how society at large is impacted. This will help see who benefits from AI and who does not.

Artificial intelligence is greatly shaping our lives in a multitude of ways. Decisions on how we work, the quality of healthcare we

DOI: 10.1201/9781003206958-1

receive, decisions on which insurance and bank loans we can get, which romantic partner we are coupled with on social media, and our consumed goods are all impacted by this seemingly mysterious abbreviation called "AI." But what is AI, and how does it impact our lives in different manners? What creates a built-in bias that can prefer person A ahead of person B when a decision is made, for example, who will be called into a job interview or who will receive an experimental drug? Or are there heterogeneous decisions made by AI based on different personal characteristics that can advantage some individuals and groups and exclude others?

AI is often treated like a *black box*, which is when

> scientific and technical work is made invisible by its own success. When a machine runs efficiently [...] one need focus only on its inputs and outputs and not on its internal complexity. Thus, paradoxically, the more science and technology succeed, the more opaque and obscure they become.
>
> (Latour, 1999, p. 304)

The black-boxing of AI means that laypeople (and even some self-proclaimed experts) have little knowledge in how, or why, AI works the way it works. This book peeks inside the "black box of AI" and contextualizes it with the society it is in, to reveal *who* are impacted by AI and in which ways—focusing on diversity matters that AI are both creating and failing at dealing with—but also how AI can be used as a tool for increased inclusion and bias mitigation. AI will be biased when the humans designing it are not aware of discriminatory actions they implicitly built into the systems and when the data AI systems are trained on do not represent the society or situation we want to create or in which the system will be used. Wittkower (2016, p. 1), for example, argues that "to minimize discriminatory effects of technical design, an actively anti-discriminatory design perspective must be adopted." AI discrimination can happen both on a group level—a company's hiring system not wanting to hire women as the company previously did not do that—and an individual level—a system targeting an individual for a low (social) credit score if the individual matches parameters associated with, for example, healthcare insurance fraud. Awareness is not enough; we need knowledge of the societal and the material transformations of AI, and knowledge, tests, and experiments that go beyond criticizing problems. AI increasingly determines our everyday lives, which calls for a reevaluation on human–machine relations (Dräger & Müller-Eiselt, 2020, p. 7).

But by only highlighting the multiple examples of AI "malfunctioning" or making bad decisions—for example, the case with *Microsoft's Twitter-bot Tay*, who after hours of learning from human Twitter users started tweeting racist tweets, for example, denying the Holocaust (Wolf et al., 2017), we get trapped in an *echo chamber* of equating AI with "bad machines". It's important to consider what ethical responsibilities developers have for the technology they create. It's also important to understand how AI can actually be used to benefit humanity and see how AI can be used as a tool to increase awareness of diversity, racial injustice, gender issues, LGBTQ+ rights, and for other marginalized or discriminated groups while simultaneously being aware of and mitigating its perils.

One example is how AI can make bosses and supervisors aware of bias in hiring and promotional processes. Several studies have shown how a job applicant's name, gender, or appearance can be used to judge them and infer their skills before the candidate has a chance to prove themselves in practice. A human bias example is how adding a headshot to one's CV is illegal, recommended, or quite random in different countries. AI, when taking into account prior biases of companies, can lay the groundwork for hiring more diverse staff (Cohen, 2019). Another example is AI that charts health risks for certain groups with major computational power, for example, by creating huge databases of health risks through image recognition and more efficiently and quickly diagnosing potential patients to provide better care (McKinney et al., 2020; Pisano, 2020; Salim et al., 2020; Wang et al., 2021). Awareness of AI and human bias enables issue mitigation. More blood samples from people aged over sixty can mitigate the bias of AI not knowing enough about the health of the elderly. More databases that include people of color can make facial recognition more applicable for nonwhites. However, there are great diversity risks of AI.

In the book, I have selected five distinct personal characteristics where I will dive into different issues of AI and diversity. I describe how AI deals with sex and gender (Chapter 2), LGBTQ+ populations (Chapter 3), race (Chapter 4), bodies, health, and aging (Chapter 5), and socioeconomic status and class (Chapter 6). I have chosen these categories as they give important examples of how AI can impact people's lives based on personal characteristics. How AI is used in connection to these illustrates deep structural issues of society on cross-cutting power dynamics of language, legislation, and labor markets. Understanding how AI impacts people in one category can tell us about discriminatory choices that relate to other categories and future technological development.

Each chapter explores and refers to research and projects directly dealing with these parameters and AI, to give a contemporary snapshot of what the world of diversity and AI looks at in the year 2023. At the end of the book, I draw on theories from science and technology studies (STS), Feminist Technoscience and intersectionality to debunk the myth of the standard human—by seeing how AI can be discriminatory when developed without proper consideration. We need responsible AI that takes issues of diversity seriously. By seeing examples of how AI works for—and against—diverse groups of the population while being aware of pitfalls, more responsible, inclusive, and robust AI systems can be created and deployed.

WHAT IS AI?

If a random person would like to know what AI is, chances are high that they will search for the term and find a dictionary. Looking up AI in dictionaries yields quite a few divergent results. The Oxford English Dictionary (2022) defines AI as "The capacity of computers or other machines to exhibit or simulate intelligent behaviour," whereas the Merriam-Webster Dictionary (2022a) defines it as "a branch of computer science dealing with the simulation of intelligent behavior in computers and the capability of a machine to imitate intelligent human behavior." Similar definitions can also be found in research, for example, in Anjila (1984, p. 65): "Artificial Intelligence is a branch of science and technology that creates intelligent machines and computer programs to perform various tasks which require human intelligence."

From these definitions, we can see a difference in, on the one hand, *simulating* humanlike intelligence and *having* humanlike intelligence. There is also the distinction between a machine having *general* intelligence or whether the machine can perform specific, *narrower*, tasks intelligently. AI is often divided into "weak AI" (AI systems that can work seemingly intelligently on a specific task) and "strong AI" (an AI system that is, per human definition, intelligent across the board). Other synonyms for strong AI vs. weak AI is general AI vs. narrow AI, which is described by Broussard (2018, p. 32) as "*General AI* is the Hollywood kind of AI…machines that think like humans. *Narrow AI* is different; it's a mathematical method for prediction… general AI is what some people want, and narrow AI is what we have." In this book, I primarily focus on existing—and thus narrow—AI.

The term AI was first coined in the summer of 1956 when the Americans John McCarthy, Marvin L. Minsky, Nathaniel Rochester, and Claude Shannon held a Summer Research Project on Artificial Intelligence in Dartmouth, NH, USA. Minsky has since been instrumental in many AI developments and discourses—for example, by collaborating on the creative design of the evil AI "HAL 9000" in the Kubrick (1968) film *2001: A Space Odyssey*. Since then, AI has been through several stages; it had an optimistic start both in terms of what was created, researched, and funded, but it also has had its down periods—themed "AI winters" (1974–1980 and 1987–1993).

AI is a subfield of *Computer Science*—the study of computers' practical and theoretical applications, involving hardware, software, and algorithms (instructions for solving specific problems and performing computations). The word computer was originally used to describe human workers who made and calculated mathematical tables (Broussard, 2018, p. 77). Computer science studies has exploded as

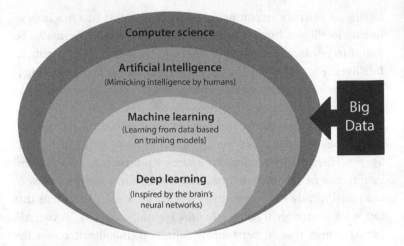

Figure 1.1 AI and related fields.

a field in the last couple of decades, both in terms of education and jobs. AI is defined as a subfield within this larger field, and it itself has several subfields, as can be seen in Figure 1.1. One of these subfields is *machine learning*, where computers learn from data, i.e., training models from datasets, which generally happens in three ways: supervised learning (where labels most often but not necessarily made by humans in the datasets guide the process), unsupervised learning (where the machine is not given guidelines), and reinforcement learning (with "punishment and rewards" for answers) (Broussard, 2018, p. 93). Under this subfield, we find *deep learning*, which is computational machine learning inspired by the human brain's networks in its complexity. Computer science and its subfields can draw on *big data*, which is huge datasets that allow computers to see patterns and find solutions based on the massive data fed into it. This book focuses on AI, but keep in mind that it is a subfield with its own subfields and that the terms used to describe all these layers of inquiry are changing over time. As this is not primarily a technical book, you will find no tutorials on programming or technical terms that require advanced degrees to understand. However, some technical expertise does help in understanding the intricate social impact of these technologies.

Artificial intelligence has been criticized as a term for describing something that is neither "artificial" nor "intelligent," for example, by Crawford (2021) in her book *Atlas of AI*. Crawford argues that AI is made from natural resources, and what makes the systems appear autonomous is people. As an example, she tells how someone who orders toilet paper through their smart voice assistant triggers a process where resources are mustered, a logistic network utilized to cut down the trees, make them into toilet paper, pack, wrap, and send it to your doorstep—as well as people involved in the logistical process, not to mention the programming and upkeep of the technologies used. Alexa, Siri, and other digital assistants are not magical witches living in a box but are made by people for people. However, the processes unleashed or magnified by AI—the natural resources consumed, who has access to the technology, and who the technology is tailored for—show that AI is not indiscriminative. Crawford argues that we must ask who is benefiting and who is excluded from power when it comes to AI. To follow this call, I explore in this book how diverse groups of people are impacted by AI, which is closely related to the concept "Augmented Intelligence" (Kurpicz-Briki, 2021; Rui, 2017; Zheng et al., 2017), which is a human-based AI approach not focusing on how AI replaced people, but rather how we can use AI to enhance and support human decision-making.

AI MACHINE LEARNING IN PRACTICE

Let us start with a thought experiment. Think of a system trained to correctly identify and separate dog and cat photos. Humans might sort some initial thousand photos with correct identifiers in a training dataset—so-called annotation of data—before the system is set to decide on new photos. Some feline commonalities might separate them from canine species, like the shape of ears, eyes, amount of fur, etc. However, this is quite the simplification, and there are multiple complex layers behind such a technology (Castelvecchi, 2016). Image recognition software does not necessarily look at pictures the way humans do. The system might work well, except when

Figure 1.2 Cat or dog?
Source: Photo from www.atchoumthecat.com

the going gets tough. Take for example the Canadian pet "Atchoum," seen in Figure 1.2, who was featured in a tweet (Twitter, 2016). Is Atchoum a dog or a cat? If you are confused, the AI sorting system might also get confused. The correct answer is that the feline fur ball has a werewolf syndrome that makes its fur grow extra quickly.

So, creatures like Atchoum might confuse humans, but what about AI? AI systems don't categorize the same ways humans do and can be confused by quite different things. One study, for example, showed highly promising accuracy when sorting between wolves and dog photos, except for when wolves were depicted during summer or dogs during winter. Why? The system had learned to recognize the snowy landscapes often associated with wolves, so instead of building a "dog OR wolf" detector, the programmers had instead built a "snow OR not" detector (Roselli et al., 2019). These animal examples might be seen as fun and goofy, but what about when humans, with our lives, rights, values, and futures are in the center of what is being fed, calculated, and interpreted by AI? To further understand that, let us look at how an AI system "decides" anything—and how this can lead to discrimination. Richardson (2021, p. 13), for example defines automated decision systems as follows:

> any systems, software, or process that use computation to aid or replace government decisions, judgments, and/or policy implementation that impact opportunities, access, liberties, rights, and/or safety. Automated Decisions Systems can involve predicting, classifying, optimizing, identifying, and/or recommending.

AI systems can make decisions that have a grander impact on a societal level, and not only for a closed dog or cat identification system. Before we jump into what happens at a societal level, I will first unwrap "being human" as not "one mold fits all" by using the concept of diversity.

WHAT IS DIVERSITY?

Diversity can be defined as "the condition of having or being composed of differing elements, especially the inclusion of people of different races, cultures, etc. in a group or organization." (Merriam-Webster Dictionary, 2022b). The concept of diversity takes into account a wide variety of personal characteristics such as gender, sexual orientation, age, race/ethnicity, physical/mental disabilities and health, socioeconomic background, and religion/spirituality, to mention some. It is, in short, what makes people different, but focused on in a positive way. Studies have shown that more diversity is crucial for work (Ilmakunnas & Ilmakunnas, 2011), education (Smith et al., 1997), knowledge production (Zhang & Guo, 2019), and health care (Salisbury & Byrd, 2006), among others.

Diversity as a concept has been calculated and has been attempted to be understood through statistical models (Simpson, 1949), economic analysis (Weitzman, 1992), and corporate attention (Hunt et al., 2015). In recent years, the focus on workforce diversity has especially emerged as a topic in HR management, company profiling, and value-driven leadership, and it has been a focus for labor scholars, for example, Tatli (2006) (*Handbook of Workplace Diversity*). Saxena (2014, p. 76) defines workforce diversity as "similarities and differences among employees in terms of age, cultural background, physical abilities and disabilities, race, religion, gender, and sexual orientation."

Diversity has however been criticized as a term for not being inclusive *enough*, i.e., "being invited to the party but not up to dance" (Cho, 2016). Diversity and inclusion programs have been operationalized across the globe but are socioculturally dependent—what works for one place does not necessarily work well in other times and places. Diversification and inclusion procedures are thus continually used, challenged, and reconceptualized. Throughout this book, although being aware of the critique of diversity as a concept, I will still use it as a starting point to help unwrap how AI impacts society and people, with an acknowledgment of the issues of the concept.

Discrimination and diversity issues are a global focus, as can be seen in the UN's 2030 Agenda for Sustainable Development, where more inclusive and equal opportunities amongst different social groups are crucial in the *Sustainable Development Goal 5 on Gender Equality, Goal 10 on Reduced Inequalities,* and *Goal 11 on Sustainable Cities and Communities.* Inequality and the fight for rights for marginalized groups are deeply interwoven in human history wherever discrimination is present. In recent years, this has been seen, for example, in the 2020 Black Lives Matter movements' protests after the killings of George Floyd by police officers—where AI has been used to identify BLM protestors in a much more precise manner than, for example, white Capitol Hill rioters in 2021. Or the 2015 European migrant crises where facial recognition debates ignited, recent debates on liberations from state control, and deepening concerns of war-purpose AI systems. These are some contemporary movements fighting for anti-discriminatory justice where AI plays a role. In short, our highly globalized world is becoming more aware of diversity issues—and in our contemporary digital societies, these movements are highly interwoven with digital technology, like social media—which plays an important role in democratic processes (Hall et al., 2018). But how does that connect with AI?

Words and language are situated and embedded in power structures, and with sensitive topics, such as inclusion and diversity, there is a risk of reproducing societal issues and provoking or using terms that become outdated quickly. But let us keep in mind that we work with terminology available to us at the given time. Although diversity as a term has its issues, there isn't an established term that works well to include the myriad of people, lives, practices, belongings, and characteristics for discussion. Similarly, "race," "class," and "gender" are highly disputed terms; some might argue that just using the terms is wrong in itself. In this book, I consciously focus on not reproducing categorical assumptions but will shine a light on how they are fueling AI's understanding of the world. Thus, we need terminology that helps us explore, for example, why we, on the one hand, have facial recognition AI that fails in general on identifying Black faces as consumers, but on the other hand identifies Black protesters well. Technology is not neutral. I implore you thus to use this book as something to spark your discussions but be aware that terminology will—and ought to—change as we progress our understandings of *socio-technical* approaches on issues of diversity in AI.

WHY AI FOR DIVERSITY—A SOCIO-TECHNICAL APPROACH

Understanding how the social world develops, deploys, and uses technological systems is a complex task, especially in regards to AI systems. Likewise, understanding how AI systems impact societies, groups, and individuals requires holistic analysis on multiple levels. AI systems are built by specific people in specific times and cultures. They can both be a tool for systematic oppression (e.g., through surveillance) and an organizational tool to help spread awareness of discriminatory actions (e.g., through hashtag optimization and sharing by political movements). To understand how society is affected by technology, and also how technology is affected back by society, a *socio-technical* approach can be useful—to understand the intricate connections between the social and technical aspects. Additionally, AI is often regarded as a "cold and objective" science, technology, engineering, and mathematics situated discipline—while diversity studies are usually found in the social science and humanities. However, interdisciplinary studies, which take knowledge and insight from multiple disciplines, are increasingly gaining importance to understand complex problems, also in more technical fields like computer programming and AI.

One of the disciplines that have investigated how technology and science are being produced is science and technology studies (STS). STS has, for many decades, argued for social construction of technology (Callon, 1987; Latour, 1987)—seeing technology not being developed and deployed in a vacuum but rather as a result of social processes and negotiations. As an STS scholar by education and training, this is the theoretical backdrop I use to analyze the topic of this book. An important first stepping stone of STS research is to open the black boxes of technology to see not only how society shapes technology but also how technology shapes society. STS has several things in common with similar fields, like *science of science* (which explore the "science behind AI") (Yuan et al., 2020).

I combine this with a gender and diversity approach from *Feminist Technoscience*—which critiques how gender and identity are described and prescribed in science and technology contexts (Barad, 2003; Haraway, 2006; Wajcman, 2007)—to understand the diversity issues that AI can bring.

In the quest for AI, it is important to be aware of social factors, for example, as Broussard (2018) describes in relation to the famous

Turing test, where two people, (A) a man and (B) a woman, are communicating in written text with a third person (C), an interrogator. After receiving the messages, C is asked to determine who is a man and who is a woman. A can lie to C, whereas B should help C. Turing then asks how replacing A with a machine would be fair if C would then be able to distinguish a human from a machine. What most people know as the Turing test is based on replacing A with a machine, and for the interrogator to then figure out who is the human and who is the machine. Broussard (2018, pp. 37–38) is not impressed by the premise:

> Turing's specifications do not conform what we now understand about gender [...] a game to determine 'intelligence,' in which the woman is assigned to be the helper? And the man is told that he can lie? The underpinnings are absurd [with] gender-coded physical and moral attributes.

Feminist Technoscience can further help us question these assumptions of technology encoded with gendered assumptions, reproductions, and values.

Let us consider the origins of AI and other social factors playing a role in how AI and computer science have developed. Had the Rockefeller Foundation not decided to fund the 1956 Summer School Research at Dartmouth or had its funding members been replaced by others, the world of computers and AI could have been different. Going even further back to 1843, had Ada Lovelace, by many considered the world's first programmer, not been hindered for being a woman (her father wanted a "glorious boy" child), or a century after that, Alan Turing not met his end in 1952 through cyanide poisoning due to homophobia, things might have played out differently. AI like any technology is not created or used independently of societal factors. This is, in STS terms, defined as "It Could Be Otherwise (or ICBO for short), allowing us to critically reflect on the world, societies and technologies not only as static predetermined variables, but rather as socially constructed" (Woolgar, 2015; Woolgar & Cooper, 1999).

Similarly, AI can construct social reality, as it both reflects and holds the power to preserve or change power relations, as well as social conditions (Ranchordás & Roznai, 2020; Richardson, 2021). This is especially visible for AI when it comes to diversity and issues of creating unbiased and inclusive technology, as we will see many examples of throughout this book. Technology is not predetermined to be used in only one way. A socio-technical perspective allows us to rethink how societies deploy and use it, opening for choices what to use, in what ways, and what not to use. Broussard (2018, p. 87) argues that "In order to create a more just technological world, we need more diverse voices at the table when we create technology," and likewise, we need technology to be made for diverse groups of the population, for technology like AI to be made more inclusive.

SOCIETAL IMPLICATIONS OF AI BIAS

AI is not created in a vacuum but has wide societal implications. Can some of these implications disadvantage and exclude certain people? Wachter-Boettcher's (2017) book *Technically Wrong: Sexist Apps, Biased Algorithms, and Other Threats of Toxic Tech* gives a broad overview of systematic discrimination and biased digital technologies, urging that it is time to "hit the break switches" and make technology companies more responsible for the technologies they make, and to get broader understandings of how technology impacts different groups of people differently.

In general, AI bias can happen in three main stages, i.e., pre, during, and after data processing. Diversity bias can, for example, occur if the data selected for a program or algorithm is not reflecting diverse enough data. An *algorithm* is a computer process that follows a set of instructions to solve specific problems. They can be quite simple, like an algorithm for sorting a random selection of numbers in increasing value, or they can be quite complex, and combined together, for example, to make high-resolution BluRay video possible. Algorithms are steps to solve a solution, for example, the instructions to solving a Rubik's cube can be considered an algorithm (Richardson, 2021,

p. 4). Selecting the best algorithm is important. If you are sorting one hundred numbers, should you go through the list first before you start sorting, or look at them one after the other, and creating the list as you go? And what if the list is not only a hundred different numbers, but rather billions of numbers and numbers that are a billion decimals long? Then, you need a plan—an algorithm—that is efficient, fast, and correct.

Sorting numbers are one of the most basic examples of algorithms in contemporary society; algorithms fuel AI decision systems that are much more complex. In complex processes, finding out what errors are or if something goes wrong can be difficult and take a lot of time and resources. On the other hand, when a "genius" algorithm is written and implemented in a system, humans can also be blind to the workings of the algorithm if it appears to be working. In the case of Tay, Microsoft's Twitter bot that became racist, and needed to be shut down the day it was released, Wolf et al. (2017, p. 1) argue that in regards to learning systems that "interact directly with people or indirectly via social media, the developer has additional ethical responsibilities beyond those of standard software. There is an additional burden of care." But what do we do if we do not see that something is wrong and take the results as great and implementable?

Data can have fair input and processing but still be misused or unfairly interpreted as output. How is that? AI systems can have a faulty presentation or focus on biased results, but the most dangerous misuse at this stage is human bias. If humans are looking for results against a specific group, systems can be rigged to produce results that disadvantage specific groups. Think back to the eugenics movement, social Darwinists, or scientific racism—most horribly realized in Nazi Germany. This was based off then-accepted scientific theories, who actively looked for "data" (in terms of statistical fabricated data) that were meant to prove that Caucasian, able-bodied, straight men were "better" than everyone else. As the age of dictatorships is far from over, we should be wary of the misuse of data, which in the age of AI and Big Data holds even greater risks. The interpretation could be gamed by powerful power-abusing leaders also in contemporary society. Richardson (2021, p. 1) warns how:

> Government agencies are increasingly using automated decision systems to aid or supplant human decision-making and policy enforcement in various sensitive social domains. They determine who will have their food subsidies terminated, how much healthcare benefits a person is entitled to, and who is likely to be a victim of a crime.

However, automated decision systems' results are often overestimated, and the full social costs are seldom known or ignored. AI systems in complex environments can lead to major trust issues if implemented too early—as an example, think of the so-called "autonomous vehicles" and self-driving tests. Infamously, several crashes have led scholars to ask if society is ready for self-driving experiments; can we accept modes of transportation being laboratories for technologies that can be life-threatening if something goes wrong, without a clear line of responsibility (Stilgoe, 2019), but in stark need of revised ethical frameworks (Evans et al., 2020).

The issue on the data output stage is primarily not a bias with AI but human bias and how we interpret data from the machine systems. AI is not biased in terms of willfully wanting to cause harm, and the systems are solving tasks with step-by-step modeling. Humans, on the other hand, are more complex. Primarily, we can sort human interpretation bias into two categories: lack of knowledge, and lack of heart. Lack of knowledge is when we do not understand the data, we trust the system, and we don't reflect in responsible manners why there could be something wrong. Ways of mitigating this are education, responsibility awareness, and having control mechanisms in place. Another out-data bias—when humans understand that computer systems are in error but still choose to interpret the data in a way that benefits them, their company, or other persuasions—is quite sinister. This is particularly dangerous, as it can be used to wrongfully justify societal policies and rules, with the justification that the system told them what was accurate or not.

SUMMARY

This chapter has given several definitions of what AI is—seeing that the term is still being negotiated by experts. Overall, AI is defined as either *simulating* or *having* humanlike intelligence. We distinguish between *general/strong* artificial intelligence, which is not yet realized, and the capacity to perform specific, *narrower/weak* tasks intelligently, which is the AI systems we have today. AI is a field within computer science, with its own subfields of machine learning and deep learning, and that it can draw on Big Data. Furthermore, I explored through a socio-technical approach how diversity—the inclusion of people of various backgrounds—can impact how AI is being built and used by society. I argued that AI as a technical artifact also constructs social realities, not free from bias, and we must open the black boxes of technology to gain a holistic understanding of how the technical and the social impact each other and dwell in symbiosis. This chapter also defined three places where bias in AI often happens: (1) pre-bias when selecting data, (2) bias when processing data, and (3) bias with out-data/processing data. Although the first two are mostly about selection difficulties and problems with how the systems are processing data, the third categorization is mostly about human bias in interpreting data. We will, in the following chapters, see several examples of how bias can occur and be accelerated in the complex socio-technical processes between humans and machines. Toward the end of the book, we will also see how diversity bias and exclusion mechanisms can be mitigated.

REFERENCES

Anjila, P. F. (1984). What is artificial intelligence? In J. Karthikeyan, Su-Hie Ting, & Yu-Jin Ng (Eds.), *Success is no accident. It is hard work, perseverance, learning, studying, sacrifice and most of all, love of what you are doing or learning to do* (p. 65). L' Ordine Nuovo Publication. https://www.researchgate.net

Barad, K. (2003). Posthumanist performativity: Toward an understanding of how matter comes to matter. *Signs: Journal of Women in Culture and Society, 28*(3), 801–831.

Broussard, M. (2018). *Artificial unintelligence: How computers misunderstand the world*. MIT Press.

Callon, M. (1987). Society in the making: The study of technology as a tool for sociological analysis. In Wiebe E. Bijker, Thomas P. Hughes, & Trevor Pinch (Eds.), *The social construction of technological systems: New directions in the sociology and history of technology* (pp. 83–103). The MIT Press.

Castelvecchi, D. (2016). Can we open the black box of AI? *Nature News*, 538(7623), 20.

Cho, J. H. (2016). *"Diversity is being invited to the party; inclusion is being asked to dance,"* Verna Myers tells Cleveland Bar. Retrieved December 6, 2021 from https://www.cleveland.com/business/2016/05/diversity_is_being_invited_to.html

Cohen, T. (2019). How to leverage artificial intelligence to meet your diversity goals. *Strategic HR Review*, 18, 62–65.

Crawford, K. (2021). *The atlas of AI*. Yale University Press.

Dräger, J., & Müller-Eiselt, R. (2020). *We humans and the intelligent machines: How algorithms shape our lives and how we can make good use of them*. Verlag Bertelsmann Stiftung.

Evans, K., de Moura, N., Chauvier, S., Chatila, R., & Dogan, E. (2020). Ethical decision making in autonomous vehicles: The AV ethics project. *Science and Engineering Ethics*, 26(6), 3285–3312.

Hall, W., Tinati, R., & Jennings, W. (2018). From Brexit to Trump: Social media's role in democracy. *Computer*, 51(1), 18–27.

Haraway, D. (2013). A manifesto for cyborgs: Science, technology, and socialist feminism in the 1980s. In Linda Nicholson (Ed.), *Feminism/Postmodernism* (pp. 190–233). Routledge.

Hunt, V., Layton, D., & Prince, S. (2015). Diversity matters. *McKinsey & Company*, 1(1), 15–29.

Ilmakunnas, P., & Ilmakunnas, S. (2011). Diversity at the workplace: whom does it benefit? *De Economist*, 159(2), 223–255.

Kubrick, S. Director (1968). *2001: A space odyssey* [Film]. Insight Media. Stanley Kubrick Productions.

Kurpicz-Briki, M. (2021). *Is augmented intelligence the AI of the future?* Retrieved March 18, 2022 from https://www.societybyte.swiss/en/2021/04/29/is-augmented-intelligence-the-ai-of-the-future/

Latour, B. (1987). *Science in action: How to follow scientists and engineers through society*. Harvard University Press.

Latour, B. (1999). *Pandora's hope: Essays on the reality of science studies*. Harvard University Press.

McKinney, S. M., Sieniek, M., Godbole, V., Godwin, J., Antropova, N., Ashrafian, H., Back, T., Chesus, M., Corrado, G. S., & Darzi, A. (2020). International evaluation of an AI system for breast cancer screening. *Nature*, 577(7788), 89–94.

Merriam-Webster-Dictionary. (2022a). *Artificial Intelligence*. Retrieved October 29, 2022 from https://www.merriam-webster.com/dictionary/artificial%20intelligence

Merriam-Webster-Dictionary. (2022b). *Diversity.* Retrieved November 29, 2021 from: https://www.merriam-webster.com/dictionary/diversity

Oxford-English-Dictionary. (2022). *Artificial Intelligence.* Retrieved November 29, 2021 from https://www.oed.com/viewdictionaryentry/Entry/271625

Pisano, E. D. (2020). AI shows promise for breast cancer screening. *Nature, 577,* 35–36. https://doi.org/10.1038/d41586-019-03822-8

Ranchordás, S., & Roznai, Y. (2020). *Time, law, and change: An interdisciplinary study.* Bloomsbury Publishing.

Richardson, R. (2021). Defining and Demystifying Automated Decision Systems. *Maryland Law Review, Forthcoming.*

Roselli, D., Matthews, J., & Talagala, N. (2019). *Managing bias in AI.* Companion Proceedings of the 2019 World Wide Web Conference. San Francisco USA.

Rui, Y. (2017). From artificial intelligence to augmented intelligence. *IEEE MultiMedia, 24*(1), 4–5.

Salim, M., Wåhlin, E., Dembrower, K., Azavedo, E., Foukakis, T., Liu, Y., Smith, K., Eklund, M., & Strand, F. (2020). External evaluation of 3 commercial artificial intelligence algorithms for independent assessment of screening mammograms. *JAMA Oncology, 6*(10), 1581–1588.

Salisbury, J., & Byrd, S. (2006). Why diversity matters in health care. *CSA Bulletin, 55*(1), 90–93.

Saxena, A. (2014). Workforce diversity: A key to improve productivity. *Procedia Economics and Finance, 11,* 76–85.

Simpson, E. H. (1949). Measurement of diversity. *Nature, 163*(4148), 688–688.

Smith, D. G., Gerbick, G. L., Figueroa, M. A., Watkins, G. H., Levitan, T., Moore, L. C., Merchant, P. A., Beliak, H. D., & Figueroa, B. (1997). *Diversity works: The emerging picture of how students benefit.* ERIC.

Stilgoe, J. (2019). Self-driving cars will take a while to get right. *Nature Machine Intelligence, 1*(5), 202–203.

Tatli, A. (2006). Handbook of workplace diversity. *Equal Opportunities International, 25*(2), 160–162. https://doi.org/10.1108/02610150610679583

Twitter. (2016). *Her.- do you have a dog or a cat? me.- I don't know.* Retrieved November 29, 2021 from https://twitter.com/1evilidiot/status/794613309613821952?ref_src=twsrc%5Etfw

Wachter-Boettcher, S. (2017). *Technically wrong: Sexist apps, biased algorithms, and other threats of toxic tech.* WW Norton & Company.

Wajcman, J. (2007). From women and technology to gendered technoscience. *Information, Community and Society, 10*(3), 287–298.

Wang, B., Jin, S., Yan, Q., Xu, H., Luo, C., Wei, L., Zhao, W., Hou, X., Ma, W., & Xu, Z. (2021). AI-assisted CT imaging analysis for COVID-19 screening: Building and deploying a medical AI system. *Applied Soft Computing, 98,* 106897.

Weitzman, M. L. (1992). On diversity. *The Quarterly Journal of Economics, 107*(2), 363–405.

Wittkower, D. E. (2016). Principles of anti-discriminatory design. 2016 IEEE International Symposium on Ethics in Engineering, Science and Technology (ETHICS). 13–14 May 2016, Vancouver, British Columbia.

Wolf, M. J., Miller, K. W., & Grodzinsky, F. S. (2017). Why we should have seen that coming: comments on Microsoft's Tay "experiment," and wider implications. The Orbit Journal, 1(2), 1–12.

Woolgar, S. (2015). It could be otherwise: provocation, irony, and limits. Center for Science, Technology, Medicine & Society at the University of California, Berkeley.

Woolgar, S., & Cooper, G. (1999). Do artefacts have ambivalence: Moses' bridges, winner's bridges and other urban legends in S&TS. Social studies of science, 29(3), 433–449.

Yuan, S., Shao, Z., Wei, X., Tang, J., Hall, W., Wang, Y., Wang, Y., & Wang, Y. (2020). Science behind AI: The evolution of trend, mobility, and collaboration. Scientometrics, 124(2), 993–1013.

Zhang, L., & Guo, H. (2019). Enabling knowledge diversity to benefit cross-functional project teams: Joint roles of knowledge leadership and transactive memory system. Information & Management, 56(8), 103156.

Zheng, N.-n., Liu, Z.-y., Ren, P.-j., Ma, Y.-q., Chen, S.-t., Yu, S.-y., Xue, J.-r., Chen, B.-d., & Wang, F.-y. (2017). Hybrid-augmented intelligence: collaboration and cognition. Frontiers of Information Technology & Electronic Engineering, 18(2), 153–179.

2

GENDERED AI

PERFORMATIVITY, EXPECTATIONS, AND SEXISM

Gender is one of the most pervasive social categories that shapes an individual's life. From the ultrasound, before one is born, to baby showers or gender reveal parties where the "pink for girls, blue for boys" binary reveals itself only to be constantly recapitulated in the toy industry. These are among the many factors that teach kids that society expects girls and boys to behave in certain ways, with little flexibility for those not fitting the gender binary. Gender is addressed by policymakers, for example, by giving preference to funding applications that have a certain percentage of women in leadership positions or, as in Norway, by requiring a certain percentage of women on corporate boards. One of the most common parameters for categorizing humans as variables in computer programs is *gender*. You might remember the old Pokémon games, where the very first decision you had to make was not which Pokémon to catch, but rather answer the binary question "Are you a boy? Or are you a girl?" which the game developers then used to determine your avatar's look. This choice only affects the visual components of the games, like what clothes are available, and some gendered dialogue

DOI: 10.1201/9781003206958-2

scenes. For humans in the real world, AI using gender as a parameter has a much wider impact. In this chapter, I provide some vantage points for understanding gender, show how AI understands gender, what sexist AI can lead to, how voice and embodied AI play with gender, and what sexbots imply for diversity.

VANTAGE POINTS FOR UNDERSTANDING GENDER

In this book, I treat gender as a social construct, meaning that "gender" is not a neutral, God-given constant, but rather a shifting and negotiated term that changes from culture to culture, and from different time periods. Take for example something as mundane as hair length. For decades, many European countries have had strict short hair for men and long hair for women, i.e., binary standards. However, Vikings considered long hair to be incredibly masculine, and in ancient China and Korea, cutting one's hair after becoming an adult was penalized. In some periods, cutting hair was seen as barbaric, antisocial, and a sign of great grief—but this changed over time and so did the social norms. Society, and hence gender, is not a static and constant entity that never changes between time and space. Simone de Beauvoir (1949) famously wrote, "One is not born, but rather becomes, a woman"—showing the social construction of the category of "woman," and that what constitutes being a woman fluctuates in time and space—an urbanite woman in Tokyo in 2022 has different expectations and social constructions than a woman from rural Moldova in the 1950s. Gender theorists such as Kristeva (1986) argue for a semiotic and symbolic understanding of gender, whereas Butler (2002) argues that gender is not something one is, but rather something one performs. Through growing up, society teaches us that women "should" behave in certain ways to reinforce femininity—but gender scholars like Butler argues that this theater of gender is socially constructed in terms of being situated in specific times, spaces, and cultures. Acting like what one thinks a certain gender should behave as thus becomes a kind of performance, reinforcing gender as a process. This gender performativity is something we also deploy on technologies we develop. A chatbot speaking in a specific pitch, the words used, and what design is implemented reproduces an understanding of what gender should mean for that technology. Gender is a complex term and hotly debated in society and among gender, feminist, queer, and other scholars. Going into all aspects of how to understand gender goes beyond the scope of this book.

Gender studies is a complex and large field with several subcategories, such as men's studies (Kimmel, 1986) and decolonial critique (Spivak, 2003), many of which are relevant for the wider topic of this book—but I have primarily chosen to focus on gender theories that directly deal with technology. In particular, I draw on the works of Perez (2019) and her book *Invisible Women: Exposing Data Bias in a World Designed for Men*, where she explains how history has a big gendered data gap; half of humanity's lives have been widely ignored throughout history—what she calls an *absent presence* when men have been taken as the norm that explains human experiences. Perez (2019) gives multiple examples and summarizes years of research on how technology discriminates against women. From office temperatures set to fit men's body temperatures to shelves stocked for male-standard heights, to car crashes that endangered women's bodies (see also Augenstein et al., 2005; Carter et al., 2014; Hitosugi et al., 2018), to health crises like heart attacks being life-threatening when women are treated with standards made from men's bodies.

For the relation between technology and gender, Feminist Technoscience scholars can give some good analytical tools, for example, Barad (2003) who argues for a post-humanist understanding of gender and intra-active entanglements between humans and nonhumans, and Haraway (1985) who with her cyborg theory argues that the biological and technical are intertwined and inseparable and that gender is also shaped by and with technology. As gender and gendering is not necessarily a binary question of "men and women," but rather quite more complex, I have also dedicated the next chapter in this book to further investigate the queering of AI bias. While there is some disagreement on how to understand gender, even within gender studies, I will focus on (social) gender as a construct that is shaped by society—not (biological) sex. I will primarily use the latter as a vantage point for exploring how gender and technology coexist in symbiosis and impact each other.

AI'S UNDERSTANDING OF GENDER

AI does not "think" in the same way as humans do but is programmed with algorithms to train models for what to do in different situations. Although some of the more advanced processes, for example, deep learning, are difficult for humans to fully grasp, let us start with an example that clearly illustrates how bias can be made—namely word vectors. Word vectors are made of words that relate to other words, for example, the word "fork" would be closer to "knife" than to "dog." Innocent, yes, but what about when humans' social lives are mixed in with the word-vectoring? Several studies on both human biases that are reproduced by machines and AI being biased itself show that it "is assumed that the nurse is a woman, and the engineer is a man" (Kurpicz-Briki & Leoni, 2021).

Let us look at another example of an algorithm that took data, processed it wrongly, and then baffled humans who had to interpret the results. From 2014 to 2018, machine learning specialists at Amazon had been working on a recruitment engine that was to sort through and select the best job applicants based on the candidates' CVs and job applications. Amazon has, as one of the world's top three largest companies, been one of the industry leaders in AI. Together, Facebook, Apple, Amazon, Netflix, Microsoft, and Google (or the FAAMNG companies for short) dominate the tech industry and global development. Comparing companies' net worth with countries, "nine companies would make their way into the top 30 wealthiest countries in the world"—and the world's eighth most profitable company—Apple—has a net worth higher than those of Russia, Italy, and Brazil combined (Lishchuk, 2021).

Thus, what the FAAMNG companies do (or fail at) in the field of AI can have a massive impact. Amazon, for example, employs over 1.3 million people globally, and who the company chooses to hire is important, also as inspiration for other companies. Hiring, as any HR manager will tell you, takes a lot of time and energy—just sifting through unqualified candidates can tie up a lot of human resources. For decades, many companies have sought to optimize the hiring process, and several systems have been put in place to attempt this. Multiple examples such as personality tests, qualification rounds, and psychological evaluations are becoming more and more mainstream, especially in the world of tech companies. Hiring the wrong person for the job can be quite costly, so getting the right person for a job is high on the priority list for employers.

When Amazon's recruitment AI was tested in 2018, it gave some results that baffled the creators (Dastin, 2018). As a product of machine learning, the Amazon recruiter tool decided that some keywords were deemed negative and should be given negative values on applications. This was mostly based on assessing the workers who currently and historically had worked for the company as the standard and trying to find workers who fit their profile best. The issue was, however, that, historically, Amazon (like most other tech companies) had mostly hired men for tech jobs. The algorithm thus regarded "woman" as a negative keyword and devalued applications that mentioned this, for example, applicants who had played in women's chess clubs or were from a women's university.

The system also favored masculine language such as "captured" and "executed," which were more often found on male engineers' CVs. Amazon ended up scrapping the idea and now uses a much more basic version, which removes duplicate applications. In this example, what was fed into the system (the 10 years of applicants, mostly men) was part of the bias. However, how it was processed and what choices the system did, could, without supervision and careful monitoring, quickly created biased exclusion. The system did not malfunction; it simply did what it was programmed to do—hire mostly as hiring had been done in the past by reifying some of the

Figure 2.1 AI hiring with bias.

norms and practices into data-driven recommendations and decisions. Figure 2.1 illustrates how it is not random who gets the job. Disruption of the *status quo* requires intervention and a will to change, otherwise, one gets stuck with what has been done before.

There are several examples of AI tools seeking to mitigate or eliminate bias in hiring processes, as described by Ramboll (2022). One example is Textio, a text editor that highlights gendered phrases in position descriptions, allowing employers to make their job advertisements more inclusive. Another is Pymetrics, an AI recruiting and job matching platform that uses personality tests or other "objective" data to allow employers to not be limited by wanting a traditionally qualified CV and thereby recruit a more diverse workforce. Other examples include unbias.io, which removes faces and names from LinkedIn profiles, as this can lead to unconscious bias when hiring, Entelo, which anonymizes interviewees by removing gender and race indicators, and Talvista, which hides characteristics and personal identifiers of the applicants and writes gender-neutral job descriptions.

SEXIST AIs?

Most AIs do not invent dialogue but have built-in human-made scripts of how they respond in certain situations. In a report by the UN from 2019, the title showcases what voice assistant Siri answers if you called her a b****: "I'd blush if I could" (West et al., 2019). Furthermore, the report provides a thorough investigation of how different voice assistant technologies like Microsoft's Cortana, Amazon's Alexa, and Apple's Siri behave. Voice assistants have ridiculed users who tell them they've been raped and need a crisis center or reacted with jokes when users say derogatory things, especially in a sexualized manner. But Cortana, Alexa, and Siri (all somewhat portrayed as *female* voice assistants) do not have the intelligence to themselves answer these questions, they have built-in queues, written by code designers. It is highly problematic if the code for answering reports of sexual assaults and toxic behavior could be written by people who think that this might be a joking matter. Similarly, problematic is the design of the bots as female. Kathleen Richardson, the author of *An Anthropology of Robots and AI*, said in an interview (Lewis, 2015) that this: "probably reflects what some men think about women—that they're not fully human beings [...] What's necessary about them can be replicated, but when it comes to more sophisticated robots, they have to be male."

In many countries, voice assistants, like Siri, have been available in a male or female voice version for a long time, but in the US, it wasn't until 2013 that Siri got a male setting. But what is it that makes these voices "female"? In countries like Japan, gendered language impacts deeply the grammar and words used, for example, the word "I" is translated to "atashi" for women, and "ore/boku" for men. In languages less impacted by gender, such as English, gendered language is primarily found in the pitch and tone of the voice (which is, of course, a simplification). Women's average voice pitch ranges from 165 to 255 Hz, whereas men's is from 85 to 155 Hz. This is a generalization based on US samples. Women in Japan, for example, use a noticeably higher pitch (Language-Log, 2017), which is partly socially dependent.

Voice assistants are normally toward mid-average, with "female" voices being lower pitched, and "male" voices being slightly higher pitched than the average. However, there are examples that break this binary, like Q, the world's first "genderless" voice assistant, according to the developers. Q is a software developed in connection to a Danish pride festival and is an experiment where a collection of nonbinary people's voices has been analyzed to provide Q with a voice pitch of 153 Hz—which was the average of the samples. This, according to the developers, represents "a future where we are no longer defined by the gender binary, but rather by our own definitions of gender, as we live and experience them" (GenderLess Voice, 2019). Q has not been widely used in the experimental field, but still provides an opportunity for critically reflecting on how and why voice assistants are gendered in the first place, and what implications this has. For sound scholars, it would be interesting to explore if this pitch (closer to the deeper than the higher pitch side) makes the technology genderless, gender-neutral, or some sort of nonbinary, as it does reproduce some societal gendered assumptions, for example, by the sample selected. At the same time, there are multiple other power dynamics when it comes to voice assistant technologies. Dialects and smaller languages are notoriously more difficult to understand for voice assistants—thus creating some discrimination against people not speaking what is considered standard and proper. Studies of languages as discriminatory are multiple, with huge implications for AI reproduction (Ferragne & Pellegrino, 2007; Roberts et al., 2014; Vicenik & Sundara, 2013; Xu et al., 2021). See also *The Smart Wife* by Strengers and Kennedy (2020) for a deeper description of the gendering of voice assistants.

Developers have a great deal of reflection to do on how they reproduce societal values. Sexism has been shown by numerous scholars and organizations to be prevalent in contemporary and historical societies, with prejudice, stereotyping, and discrimination against women limiting the opportunities and life experiences of women across the globe. Sexism across the globe takes many forms, including threatening behavior, exclusion, rigid roles, sexualization, shaming, catcalling, sexist language, insults, control, harassment, stalking, physical and psychological violence, rape, murder, and deprecation of women (Human-Rights-Channel.coe, n.d.). Some attempts to stop this through the use of AI include work on ending men's violence toward women by detecting violent behavior from video, for example, through support vector machine approaches to detect from video surveillance if the behavior is abusive (Ramboll, 2022, p. 17).

But digital technologies can also risk reproducing such sexism, and we need to carefully reflect on the connections between the social and the technical. This can also be on levels that might not seem as serious at first glance. One such reflection is how and why digital assistants are portrayed in a female manner, to begin with. Microsoft's Cortana is, for example, based on the sexualized Cortana character from the *Halo* game series, a nude-looking blue woman wearing only a minimalistic swimsuit. Microsoft has since tried to distance the Cortana voice assistant from the sexualized game character. Another example of technology reproducing sexism is the now infamous pitch for "TitStare," an app for men who look at women's breasts (Broussard, 2018, p. 167; Ohlheiser, 2013). Broussard (2018, p. 85) describes this issue as follows:

> we have a small, elite group of men who tend to overestimate their mathematical abilities, who have systematically excluded women and people of color in favor of machines for centuries, who tend to want to make science fiction real, who have little regard for social convention, who don't believe that social norms or rules apply to them, who have unused piles of government money sitting around, and who have adopted the ideological rhetoric of far-right libertarian anarcho-capitalists. What could possibly go wrong?

Quite a lot can go wrong when the imagined user is similar to homogenous groups of developers when "I-methodology" is used in design (Oudshoorn et al., 2004). Configuring the user as "I myself," for example, if young white straight guys imagine their users to also be just that, users who don't fit the bill might be quite excluded from whatever is developed—a "the default male" (Irigaray, 1992; Perez, 2019). However, this default male has historically led to an enormous amount of sexism. It is well worth critically examining how new pathways that can eliminate such gendered practices and infrastructures can be mitigated instead of reinforced by digital technology such as AI. Take for example Austria's employment agency, which used a hiring algorithm that graded women and disabled people lower than men and able-bodied individuals, devalued people over 30 years of age, and gave mothers lower scores than fathers (dig.watch, 2019). Here, being a woman is intersecting with other personal parameters as well, such as raising children. There is a reason why it is illegal in many countries to ask applications whether or not they are pregnant or have children, as this can be devalued by employers—which then, as we see in Austria, is reproduced by AI. In a study of this, Allhutter et al. (2020) describe that "the design of the algorithm is influenced by technical affordances, but also by social values, norms, and goals," with scholars calling for transparency, the right to appeal, and a need for public participation (Mager & Allhutter, 2021). As data is never neutral, it is worth questioning what values and norms are built into the technology. One study that uses AI to identify gender-based discrimination in oral settings is the gender-identifying open-source speaker detection, using French radio as a case to raise awareness of the amount of speaking time women vs. men have (Doukhan et al., 2018), and a Swedish solution for evaluating the outcome of diversity and inclusion initiatives taken at a workplace called Ceretai.

ROBOTS AND GENDER

Up until now, this book has primarily discussed software and programs that have a minimal physical presence, for example, in a smart speaker. But let us also consider physical containers of AI, aka *hardware*. Robots are increasingly becoming part of people's lives, including service robots such as vacuum cleaner robots and lawn-mower robots. Also, social robots such as Pepper and Sofia the Robot are making their way into research, industry, and the public's techno-logical zeitgeist. The robots have built-in AI systems that interact with their physical components. Robots, with varying complexity of AI, are often portrayed in a gendered manner. This is especially so for robots taking on a female depiction, like Sofia the robot, who is modeled after Audrey Hepburn, Queen Nefertiti of ancient Egypt, and the creator's wife.

In "Mechanical genders: How do humans gender robots?" I discuss how the more humanoid a robot becomes the more we humans tend to gender it (Søraa, 2017). Androids, i.e., robots anthropomorphized to look like humans, are particularly interesting in a gendered con-text. A robot does not "have a gender" as we humans understand the term as an adjective or noun, but we humans *gender* it as a verb, partly through the robot's design, and partly through societal cues and fac-tors where it is implemented. When the designers of Sofia the robot chose to model "her" after Audrey Hepburn, and give the impression of the robot wearing lipstick and long lashes, it "performs" certain femininity (Søraa & Bruijning, forthcoming). When the creators of Pepper (or at least their social media theme) describe Pepper's lack of gender on their website, that also matters (Softbank, n.d.):

Pepper is a robot that has no gender.

WHY? Pepper is neither male nor female, but as you get to know Pepper, don't be surprised if you find yourself referring to Pepper in a gender that makes the most sense to you.

HOW? Pepper should be viewed as just a robot, a friendly one.

Pepper is a humanoid, but not a human. Pepper is a robot that has no gender.

However, robots are assigned a societal gender the more humanoid they become. We humans have a strong need to categorize and place things in our social world. Language is gendered (some languages more so than others), is it "la Pepper, le/el/il Pepper" in Romance Languages such as French, Spanish, and Italian? Is it "han Pepper" or "hun Pepper" in Norwegian gendered dialects? Describing Pepper in English, do we refer to it, he, or she? In Japanese, does it use masculine "boku," feminine "atashi," or the more neutral "watashi" to refer to itself? The linguistic choices the speaker/writer makes are part of the gendering of robots, which shows how the categorizations through language portray and reproduce societal values, which are also translated to data and materialities (Figure 2.2).

One can also observe how clothes for robots are quite gendered. When Pepper is dressed up as *Little Red Riding Hood*, it tells a (gendered) story—and technologized stories matter. I am not here to judge whether or not gendering robots is a good or bad thing (as with binary philosophical questions, the answer is often more complex), but my point is that we need to be aware that gendering of technology

Figure 2.2 Robots performing gender.

is a field that has been studied for a long time, with a focus on many technologies, including the microwave oven (Cockburn & Ormrod, 1993), electric shavers and bicycles (Oudshoorn et al., 2002), and telephones (Frissen, 2018). Robots and AI are not, nor should they be, an exception to this inquiry. However, the novelty I would argue with reading robots in a *Feminist Technoscience* tradition is the humanness replication of the technology. How we should understand, discuss, and govern robots are strongly debated. Scholars like Bryson (2010, p. 1) warn against humanizing robots, as it could imply a dehuman-ization of people: "Robots should not be described as persons, nor given legal nor moral responsibility for their actions. Robots are fully owned by us. We determine their goals and behavior, either directly or indirectly through specifying their intelligence."

AI AND SEX

Technology has always had a strong connection to the sex indus-try, and AI is no exception. From realistic sexbots—although with still limited conversation options—to customized sex tools that learn from their users, with names like "autoblow," "HUM," "Smart lipstick," and "VR fantasies," AI is looking at the sex industry as a huge commercial market. One of the more troubling cases is how some porn videos are being made with "Deep-fake" technologies, which put the faces of, for example, celebrities onto porn actors to give illusions (Maddocks, 2020). This is problematic from a privacy issue, as well as from an ethical issue, and can hurt the careers of actors if mistaken for real videos—let alone ordinary people with vengeful exes in times of "revenge-porn"—or "image-based sexual abuse," as some academics have suggested calling it (McGlynn & Rackley, 2017).

In related notions, sex robots have also faced much criticism and scholarly interest in recent years. Danaher (2017) suggests three main ethical concerns on sex robots: "(i) Do they benefit/harm the user? (ii) Do they benefit/harm society? or (iii) Do they benefit/ harm the robot?" In a paper called "Are we ready for sex robots?"

Scheutz and Arnold (2016) discuss how men are more inclined to find them appropriate, but women are less so. Appel et al. (2019) connect sex robots to fandom or *otaku* culture in Japan, whereas Richardson (2016) discusses the asymmetrical "relationship" parallels between prostitution and the development of sex, seeing how anthropomorphism reflects gendered notions of sexuality. One conundrum in terms of AI is how these sex robots are *learning* from users and building profiles based on likes and dislikes. Some scholars suggest that sex robots could be important tools for disabled people or people who would otherwise struggle with finding human sexual partners (Devlin, 2015; Jecker, 2021). Sex robots are an example of technology on a charged topic, as sexuality has often been throughout human history.

SUMMARY

This chapter explored the diversity criteria of "gender"—which I first unwrapped to see a wide variety in how gender is conceptualized. AI uses gender as a parameter to categorize humans and nonhumans, and by doing so reproduces certain societal norms and values we have about gender. But these understandings of what gender is, or should be, are not constant, but socially constructed through their performativity. We saw an infamous example of Amazon's hiring system sorting through job applications, where women applicants were devalued as their applications did not fit the profiles of most people already working at Amazon. Further, I looked at sexist apps, with Siri, Alexa, and other app sisters responding poorly to cries for help after serious sexual assaults and flirting back when human users sexualized the apps. I had a look at the voice assistant Q which is pioneering gender-neutral voice assistance, and on the physical hardware side, I looked at how robots are being gendered in the process of anthropomorphizing. Finally, AI for sex applications was discussed, with the problematization of deep-fake porn videos. AI has little understanding of the deep societal constructions that gender means for societies, and how gender is understood differently across times and cultures to tell. We must thus be careful what norms we think are norm(al)s today, and careful for thinking that how one society and group defines gender is universal. To further unwrap gender as something nonbinary, the next chapter looks at how AI can be queered.

REFERENCES

Allhutter, D., Cech, F., Fischer, F., Grill, G., & Mager, A. (2020). Algorithmic profiling of job seekers in Austria: how austerity politics are made effective. *Frontiers in Big Data, 3*, 5.

Appel, M., Marker, C., & Mara, M. (2019). Otakuism and the appeal of sex robots. *Frontiers in Psychology, 10*, 569.

Augenstein, J., Perdeck, E., Bahouth, G. T., Digges, K. H., Borchers, N., & Baur, P. (2005). Injury identification: Priorities for data transmitted. *Proceedings of the 19th International Technical Conference on the Enhanced Safety of Vehicles (ESV)*. 6–9 June, 2005, Washington D.C.

Barad, K. (2003). Posthumanist performativity: Toward an understanding of how matter comes to matter. *Signs: Journal of Women in Culture and Society, 28*(3), 801–831.

Broussard, M. (2018). *Artificial unintelligence: How computers misunderstand the world*. MIT Press.

Bryson, J. J. (2010). Robots should be slaves. In Yorick Wilks (Ed.), *Close engagements with artificial companions: Key social, psychological, ethical and design issues* (pp. 63–74). John Benjamins Publishing Company.

Butler, J. (2002). *Gender trouble*. Routledge.

Carter, P. M., Flannagan, C. A., Reed, M. P., Cunningham, R. M., & Rupp, J. D. (2014). Comparing the effects of age, BMI and gender on severe injury (AIS 3+) in motor-vehicle crashes. *Accident Analysis & Prevention, 72*, 146–160.

Cockburn, C., & Ormrod, S. (1993). *Gender and technology in the making*. SAGE Publications.

Danaher, J. (2017). Should we be thinking about sex robots? In John Danaher, & Neil McArthur (Eds.), *Robot Sex: Social Implications and Ethical* (pp. 3–14). MIT Press.

Dastin, J. (2018). *Amazon scraps secret AI recruiting tool that showed bias against women*. Reuters.com. Retrieved December 6, 2021 from https://www.reuters.com/article/us-amazon-com-jobs-automation-insight-idUSKCN1MK08G

de Beauvoir, S. (1949). *Le Deuxieme Sexe, Vol 1: Les Faits Et Les Mythes*.

Devlin, K. (2015). In defence of sex machines: Why trying to ban sex robots is wrong. The Conversation.

dig.watch. (2019). Discriminatory employment algorithm towards women and disabled. Retrieved February 26, 2022 from https://dig.watch/updates/discriminatory-employment-algorithm-towards-women-and-disabled

Doukhan, D., Carrive, J., Vallet, F., Larcher, A., & Meignier, S. (2018). An open-source speaker gender detection framework for monitoring gender equality. 2018 IEEE International Conference on Acoustics, Speech and Signal Processing (ICASSP). 15–20 April 2018, Calgary, Alberta, Canada.

Ferragne, E., & Pellegrino, F. (2007). Automatic dialect identification: A study of British English. In C. Müller (Eds.), Speaker Classification II. Lecture Notes in Computer Science (vol 4441, pp. 243–257). Springer. https://doi.org/10.1007/978-3-540-74122-0_19

Frissen, V. (2018). Gender is calling: Some reflections on past, present and future uses of the telephone. Taylor & Francis.

GenderLess Voice. (2019). Meet Q: The first genderless voice – full speech. Retrieved December 6, 2021 from https://www.genderlessvoice.com/

Haraway, D. J. (1985). A manifesto for cyborgs: Science, technology, and socialist feminism in the 1980s. Center for Social Research and Education.

Hitosugi, M., Koseki, T., Hariya, T., Maeda, G., Moriguchi, S., & Hiraizumi, S. (2018). Shorter pregnant women restrained in the rear seat of a car are at risk for serious neck injuries: Biomechanical analysis using a pregnant crash test dummy. Forensic Science International, 291, 133–137.

Human-Rights-Channel.coe. (n.d.). Sexism: See it. Name it. Stop it. Council of Europe. Retrieved February 26, 2022 from https://human-rights-channel.coe.int/stop-sexism-en.html

Irigaray, L. (1992). Elemental passions. (J. Collie & J. Still, Trans.). Athlone.

Jecker, N. S. (2021). Nothing to be ashamed of: Sex robots for older adults with disabilities. Journal of Medical Ethics, 47(1), 26–32.

Kimmel, M. S. (1986). Introduction: Toward men's studies. American Behavioral Scientist, 29(5), 517–529.

Kristeva, J. (1986). The kristeva reader. Columbia University Press.

Kurpicz-Briki, M., & Leoni, T. (2021). A world full of stereotypes? Further investigation on origin and gender bias in multi-lingual word embeddings. Frontiers in Big Data, 4, 20.

Language-Log. (2017). Nationality, gender and pitch. Retrieved December 6, 2021 from http://itre.cis.upenn.edu/~myl/languagelog/archives/005104.html

Lewis, T. (2015). Rise of the fembots: Why artificial intelligence is often female. https://www.livescience.com/49882-why-robots-female.html

Lishchuk, R. (2021). How large would tech companies be if they were countries? Retrieved December 6, 2021 from https://mackeeper.com/blog/tech-giants-as-countries/

Maddocks, S. (2020). 'A deepfake porn plot intended to Silence me': Exploring continuities between pornographic and 'political'deep fakes. *Porn Studies, 7*(4), 415–423.

Mager, A., & Allhutter, D. (2021). *How fair is the AMS algorithm?* https://epub.oeaw.ac.at/?arp=0x003c4bd9

McGlynn, C., & Rackley, E. (2017). Image-based sexual abuse. *Oxford Journal of Legal Studies, 37*(3), 534–561.

Ohlheiser, A. (2013). *Meet 'Titstare,' the tech world's latest 'joke' from the minds of brogrammers.* TheAtlantic. Retrieved December 6, 2021 from https://www.theatlantic.com/technology/archive/2013/09/titstare-tech-worlds-latest-brogrammer-joke-techcrunch-disrupt/311308/

Oudshoorn, N., Rommes, E., & Stienstra, M. (2004). Configuring the user as everybody: Gender and design cultures in information and communication technologies. *Science, Technology, & Human Values, 29*(1), 30–63.

Oudshoorn, N., Saetnan, A. R., & Lie, M. (2002, July). On gender and things: Reflections on an exhibition on gendered artifacts. *Women's Studies International Forum.* (Vol. 25, No. 4, pp. 471–483). Pergamon.

Perez, C. C. (2019). *Invisible women: Exposing data bias in a world designed for men.* Random House.

Ramboll. (2022). *AI for gender equality—Addressing inequality through AI.* https://www.vinnova.se/globalassets/mikrosajter/ai-for-jamstalldhet-starker-tillvaxten-samhallsekonomin-och-arbetsmarknaden/ramboll---ai-for-gender-equality-2020-11-19.pdf

Richardson, K. (2016). The asymmetrical 'relationship' parallels between prostitution and the development of sex robots. *Acm Sigcas Computers and Society, 45*(3), 290–293.

Roberts, C., Davies, E., & Jupp, T. (2014). *Language and discrimination.* Routledge.

Scheutz, M., & Arnold, T. (2016). *Are we ready for sex robots?* 2016 11th ACM/IEEE International Conference on Human-Robot Interaction (HRI). 7–10 March, Christchurch New Zealand.

Softbank. (n.d.). *Pepper is a humanoid-robot.* Retrieved September 18, 2022 from https://developer.softbankrobotics.com/pepper-qisdk/design/pepper-humanoid-robot

Søraa, R. A. (2017). Mechanical genders: How do humans gender robots? *Gender, Technology and Development, 21,* 99–115.

Søraa, R. A., & Bruijning, N. (forthcoming). Gendering Sophia the robot: Cyborg, fembot, fashionista, citizen, imagination, and boundary object. In NA (Ed.), *Gender of things.* Routledge.

Spivak, G. C. (2003). Can the subaltern speak? *Die Philosophin, 14*(27), 42–58.

Strengers, Y., & Kennedy, J. (2020). *The smart wife: Why Siri, Alexa, and other smart home devices need a feminist reboot.* Mit Press.

Vicenik, C., & Sundara, M. (2013). The role of intonation in language and dialect discrimination by adults. *Journal of Phonetics*, 41(5), 297–306.

West, M., Kraut, R., & Ei Chew, H. (2019). *I'd blush if I could: closing gender divides in digital skills through education.* https://unesdoc.unesco.org/ark:/48223/pf0000367416

Xu, F., Dan, Y., Yan, K., Ma, Y., & Wang, M. (2021). Low-resource language discrimination toward chinese dialects with transfer learning and data augmentation. *Transactions on Asian and Low-Resource Language Information Processing*, 21(2), 1–21.

3

QUEERING AI

GENDER EXPRESSION, IDENTITY, AND BINARIES

Gender diversity is more than the binary counting of women and men like many computer programs like to "count gender diversity." To further unwrap gender, I have chosen to focus particularly on sexual orientation and gender expression, through a queer perspective, by focusing on how members of the LGBTQ+ community are impacted by—and impacts—AI. LGBTQ+ stands for Lesbian, Gay, Bisexual, Transgender, Queer, plus any others who identify within the community and is an umbrella term that is disputed and culturally contextualized. I will, in this book, use the term "queer" to refer to this minority group while recognizing that it both includes and excludes, and whilst other acronyms and categorizations have been suggested, queer as a term can serve as a vantage point from which to consider how AI is dealing with gender and sexuality, opening our understanding beyond nonbinary, which ironically challenges the computer binary 0–1 built-up structure of logical thinking.

DOI: 10.1201/9781003206958-3

AI'S POTENTIAL FOR DISCRIMINATION

Guessing users' sexuality has a wide array of impacts. From the companies' perspectives, it makes sense in marketing and adds a revenue point of view, where learning more about the individual user will provide insight into what that user is more likely to click on. Knowing if a user is more likely to click on an ad for yarn than drag makeup allows for more targeted advertising. However, there is a grimmer side to the picture. Sexual orientation is for many people across the world a highly personal matter, especially in regions with controlling states and hostile communities. It is important to not Westernize being LGBTQ+, but ask which queer people, where? Being outed as queer can have a major impact on individuals who do not have legal, community, or financial support to live their lives in places where being gay is criminalized, or exclude people from access to basic rights, healthcare, and work, and can even lead to death. Queer people are globally facing widespread discrimination, and while many countries are much more open to queer people living their lives than just some decades ago, many places across the globe have severe laws against same-sex couples, from legal discrimination in multiple countries. Being gay is still punishable by death in several countries.

For the queer inhabitants of these nations, who without a shred of doubt exist and are living under life-threatening conditions just because of who they are. AI such as social media predictions or facial inferring of sexuality can mean a death sentence if used in the wrong hands. Within these countries, any technology that allows the governing forces to abuse their power to surveil, control, penalize, and even kill off their inhabitants is at great risk of human rights violations. AI is thus not only a question about technology—it is highly political—and its design can for certain individuals be the difference between life and death. For a deeper dive into how AI can be regulated in a trustworthy manner, see the "Queer in AI Approach to Artificial Intelligence Risk Management" (Agnew et al., 2021).

Historically, queer people have globally been on a trajectory to more rights. In April 1952, the American Psychiatric Association listed homosexuality as a sociopathic personality disturbance, and the year after, then American President Eisenhower signed an executive order that "banned homosexuals" from working for the federal government. The 1969 Stonewall protests, where police cracked down on the queer community in New York, are often recognized as the beginning of the queer civil rights movement in the United States. However, setting 1969 as the beginning of the movement does obscure earlier efforts, for example, the Mattachine Society, a gay rights organization founded in 1950, or the Compton Cafeteria riots in 1966 in San Francisco, where the transgender community also fought against police harassment and violence. On December 15, 1973, the American Psychiatric Association removed homosexuality from its list of mental disorders, and globally, the World Health Organization did the same on May 17, 1990. On March 2, 1982, Wisconsin became the first state to outlaw discrimination based on sexual orientation. Gay marriage was legalized in the Netherlands on April 1, 2001, followed by most other Western democracies in the next decades, with the UN passing the first-ever Gay Rights Protection Resolution in June 2011.

However, being gay is still not globally recognized as having human rights. In October 2009, a Ugandan MP introduced a "Kill the Gays" bill. At the height of the COVID-19 pandemic, queer people were scapegoated in Ukraine and Senegal for the divine punishment believed to have caused the pandemic, and in Panama, transgender people were harassed no matter which of the binary "men or women" quarantine days they ventured outside (Reid, 2020). At the same time, scholars like Puar (2013) urge us to rethink "homonationalism" as something not only relating to legal rights, questioning the deeper societal processes that make "gay friendly" a measurement of civilized countries.

Why does this matter for AI and diversity? How societies treat minority groups should matter a whole lot to you, even if you are not in the LGBTQ+ community. As history time and again shows, majority discrimination against vulnerable groups and minorities does happen, and heavily impacts the quality of life of those discriminated against. Society at large benefits from diversity and inclusion—and care for all is a core human value. The next time it could be you who is scrutinized for some aspect of your being. Remember the old practice of bringing canaries into the mines when digging for minerals? The birds would sign as long as oxygen levels were sufficient but would grow silent when oxygen conditions worsened. We need to be aware of the digital canaries, those minorities that first see exclusion and discrimination, as grave warnings of what to mitigate, eliminate, and ensure does not grow into larger societal AI-based control and discrimination. LGBTQ+ scientists, especially in STEM fields, struggle with "heteronormative assumptions can still create less conscious forms of bias, and an unwelcoming environment that puts scientists from sexual and gender minorities (LGBTQ) at a disadvantage" (Freeman, 2018). With AI technology, control and discrimination are very treacherous if not implemented in responsible ways. Let us consider the technologies that have been shown to have an impact on this.

PREDICTING GENDER AND SEXUALITY

What we press "like" for, what we "follow," and what we "share" on social media tell a lot about who we are as people—which can be learned and utilized by AI systems. A study by Kosinski et al. (2013) from Cambridge University that used algorithms to predict the religion, politics, race, and sexual orientation of about 58,000 Facebook users correctly assessed the distinction "between homosexual and heterosexual men in 88% of cases, African Americans and Caucasian Americans in 95% of cases, and between Democrat and Republican in 85% of cases." Is this a simple, harmless, guessing game, or could it lead to discriminatory action? On social media, guessing gender is big business as it connects to marketing revenue, and can get muddy when tied to sexuality. Several users have reported how the social media algorithms knew they were gay or bisexual before they realized it themselves, as we can read in this case of TikTok inferring a previously identified straight woman:

> There is something about TikTok that feels particularly suited to these journeys of sexual self-discovery and, in the case of women loving women, I don't think it's just the prescient algorithm. The short-form video format lends itself to lightning bolt-like jolts of soul-bearing nakedness, with the POV camera angles bucking conventions of the male gaze, which entrenches the language of film and TV in heterosexual male desire.
>
> (Joho, 2021)

Even if social media can be right—whether the user is ready to hear it or not—it can also be wrong. As AI is a key component of social media, the algorithms used should be paid close attention to. A study from Fosch-Villaronga et al. (2020) showed, for example, that gay men were misgendered by gender inverting on social media platform Twitter much more often than straight men and that nonbinary people were reported to be misgendered in an overwhelming amount. That is due to the majority categorization, of dividing what you click, like, and comment on to fit within what Twitter deems to be "feminine" or "masculine" interests, and then categorizing you as a user as either male or female based on your interests, like Figure 3.1 illustrates.

But this binary distinction into "men are masculine" and "women are feminine" further stereotypes the "Men are from Mars; Women are from Venus," which creates two categories of humans almost living on different planets with all the constraints that imply. If not critically intervened, boys grow up in a blue world of commodification of toys and interests to facilitate tough masculinity with pirates, starships, and weapons, whereas girls are offered a pink-washed fairy tale of sparkles, unicorns, and housework. Societal pressure on what gender expression and interests are, is enormous, and this is the historic data that machines risk reproducing.

Figure 3.1 Gender inferring on social media.

CATEGORIZING AND EXCLUSION

Queer communities often have a fluidity of gender understanding, smashing sexual and gender binaries. Remember from the previous chapter on gender, how we are often taught to perform a certain gender role, for example, a gay guy told to "man up" and play sports instead of dance, or to date girls, as that is what traditional societal gender roles proclaim is "correct." But situated in our native cultures, we often forget that what goes for being "masculine" or "feminine" in our own subcommunity might be completely alien to other places in the world. The issue with computers is that they can be flatly incapable of working beyond the binary of male and female regarding gender identity, especially for surveillance technologies (Katyal & Jung, 2021). It has, for example, been reported that when AI systems are used to monitor inappropriate social media content, they disproportionately flag and remove content from transgender, conservative, and Black social media users. While politically conservative content was generally removed for violating actual site guidelines, transgender and Black users' content was shown removed due to being flagged as "inappropriate" despite complying with site policies (Haimson et al., 2021). Bot networks—hijacked computers infected by malware—have been reported to negatively target transgender users, as well as "hate in the machine" toward Black and Muslim users (Williams et al., 2020). False information and bot-takeover of topics are a systematic problem on social media and were also seen during the COVID-19 pandemic (Himelein-Wachowiak et al., 2021).

One issue is AI's inability to understand subcultures. A study of famous Twitter users and how toxic their accounts and tweets were deemed by the AI technology of Jigsaw (owned by Alphabet, who owns Google) looked at how drag queens' who used group-internal slang like the friendly "Hey bitch, I love you, how are you doing" compared to more white nationalists and right-wing politicians, who tweeted racist messages but in a convoluted way (Oliva et al., 2021). Toxicity was here seen as rude and disrespectful comments that are likely to make people leave a discussion. The system was trained by asking people to rate comments. They found that "the toxicity levels of the drag queens' accounts ranged from 16.68 percent to 37.81 percent, while the white nationalists' averages spanned from 21.30 percent to 28.87 percent." One of the issues discovered was that the algorithm tagged specific words as toxic, like "gay," "lesbian," and "queer"—leading gay people tweeting about their queer life to automatically be tagged as quite toxic (Antonialli, 2019). Counterexamples also exist, like the Perspective API, which can evaluate posts and comments for cyberbullying or harassment using machine learning techniques (Ramboll, 2022, p. 17).

In the case of drag queens, who often reclaim slurs in their sharp-tongued communication style—and invoke that in their personal brand as entertainers—the risk is that their tweets can be silenced, whereas others, like white nationalists, are deemed much less toxic, even though some of them got eventually banned from Twitter and other platforms. Banning certain individuals with hostile opinions who condone violence on social media platforms is one step to take to mitigate unwanted behavior. Social media manipulation is big business, both on a state level (e.g., as with foreign intervention in American elections), for extortion and terrorism, and for private company interests. Flagging LGBTQ+ content as adult even when it complies with site guidelines is just one example. "Shadowbanning"—making content or users less available for certain areas or communities—is another example of how social media is manipulated, both from social media companies themselves, for example, Twitter limiting the reach of some individual posts by politicians or others that are contrary to their policy on COVID-19 or election conspiracy disinformation. However, there are also several examples of how trans lives can gain more visibility and peer encouragement on social media, such as YouTube communities (O'Neill, 2014; Raun, 2016; Tortajada et al., 2021), where visibility and outreach are key.

Another study on "LGBTQ-AI? Exploring Expressions of Gender and Sexual Orientation in Chatbots" by Edwards et al. (2021) found that although "chatbots are proficient in using language to express identity, they also display a lack of authentic experiences of gender and sexuality." So, although chatbots can mimic some gendered understandings of the world, this performativity does not imply that they have a deeper understanding of how the world works—especially in such a complex sociocultural phenomenon as gender and diversity.

SUMMARY

In this chapter we have explored how gender diversity is more than the binary counting of women and men. It is a complex topic that consists of gender identity, sexuality, societal gendering norms, and more. Although some programs try to determine users' sexuality or gender identity, this is a dangerous path, as it paves the way for exclusion and bias. When it comes to gender identity, the potential for harm is big, as this has historically been a part of people's lives that has been widely discriminated upon. Thus, as the chapter described, predicting and categorizing people based on for example their sexuality can lead to harmful results. We saw for example how social media content was flagged as inappropriate, and how chatbots lack an understanding of this topic. We also saw how drag queens were tagged as proclaimers of hate speech, over extremists. Although computer programs are good at categorizing broadly, not all human characteristics should necessarily be counted.

REFERENCES

Agnew, W., Pajaro, J., & Subramonian, A. (2021). Rebuilding trust: Queer in AI approach to artificial intelligence risk management. *arXiv preprint arXiv:2110.09271*

Antonialli, D. (2019). Drag queen vs. david duke: Whose tweets are more 'toxic'. *Wired, July/August edition.* Retrieved September 20, 2022 from https://www.wired.com/story/drag-queens-vs-far-right-toxic-tweets/

Edwards, J., Clark, L., & Perrone, A. (2021). LGBTQ-AI? Exploring expressions of gender and sexual orientation in chatbots. *CUI '21: Proceedings of the 3rd Conference on Conversational User Interfaces.* 27–29 July 2021, Spain. ISBN: 978-1-4503-8998-3.

Fosch-Villaronga, E., Poulsen, A., Søraa, R. A., & Custers, B. H. (2020). Don't guess my gender gurl: The inadvertent impact of gender inferences. *BIAS 2020: Bias and Fairness in AI Workshop at the European Conference on Machine Learning and Principles and Practice of Knowledge Discovery in Databases (ECML-PKDD)* (pp. 1–9). 14–18 September 2020.

Freeman, J. (2018). LGBTQ scientists are still left out. *Nature, 559*, 27–28. https://doi.org/10.1038/d41586-018-05587-y

Haimson, O. L., Delmonaco, D., Nie, P., & Wegner, A. (2021). Disproportionate removals and differing content moderation experiences for conservative, transgender, and black social media users: Marginalization and moderation gray Areas. *Proceedings of the ACM on Human-Computer Interaction, 5*(CSCW2), 1–35.

Himelein-Wachowiak, M., Giorgi, S., Devoto, A., Rahman, M., Ungar, L., Schwartz, H. A., Epstein, D. H., Leggio, L., & Curtis, B. (2021). Bots and misinformation spread on social media: Implications for COVID-19. *Journal of Medical Internet Research, 23*(5), e26933.

Joho, J. (2021). *TikTok's algorithms knew I was bi before I did. I'm not the only one.* Mashable. Retrieved February 26, 2022 from https://mashable.com/article/bisexuality-queer-tiktok

Katyal, S., & Jung, J. (2021). The gender panopticon: Artificial intelligence, gender, and design justice. *UCLA Law Review, 68*, 692.

Kosinski, M., Stillwell, D., & Graepel, T. (2013). Private traits and attributes are predictable from digital records of human behavior. *Proceedings of the National Academy of Sciences*, 110(15), 5802–5805.

O'Neill, M. G. (2014). Transgender youth and YouTube videos: Self-representation and five identifiable trans youth narratives. In Christopher Pullen (Ed.), *Queer youth and media cultures* (pp. 34–45). Springer.

Oliva, T. D., Antonialli, D. M., & Gomes, A. (2021). Fighting hate speech, silencing drag queens? artificial intelligence in content moderation and risks to LGBTQ voices online. *Sexuality & Culture*, 25(2), 700–732.

Puar, J. (2013). Rethinking homonationalism. *International Journal of Middle East Studies*, 45(2), 336–339.

Ramboll. (2022). *AI for gender equality—Addressing inequality through AI*. https://www.vinnova.se/globalassets/mikrosajter/ai-for-jamstalldhet-starker-tillvaxten-samhallsekonomin-och-arbetsmarknaden/ramboll---ai-for-gender-equality-2020-11-19.pdf

Raun, T. (2016). *Out online: Trans self-representation and community building on YouTube*. Routledge.

Reid, G. (2020). *A global report card on LGBTQ+ Rights for IDAHOBIT*. Retrieved February 22, 2022 from https://www.hrw.org/news/2020/05/18/global-report-card-lgbtq-rights-idahobit

Tortajada, I., Willem, C., Platero Mendez, R. L., & Araüna, N. (2021). Lost in transition? Digital trans activism on Youtube. *Information, Communication & Society*, 24(8), 1091–1107.

Williams, M. L., Burnap, P., Javed, A., Liu, H., & Ozalp, S. (2020). Hate in the machine: Anti-Black and anti-Muslim social media posts as predictors of offline racially and religiously aggravated crime. *The British Journal of Criminology*, 60(1), 93–117.

4

AI AND RACE

RECOGNITION, BIAS, AND SYSTEMIC ISSUES

"Race" is one of the most contested and discussed terms in human history—and multiple studies have shown that societies across the globe struggle with racial injustice, unfairness, and human-biased discrimination (Appiah, 2010; Griffin, 2012; Miles, 2004; West, 2017). Racism is defined by the Oxford Dictionary as: "prejudice, discrimination, or antagonism by an individual, community, or institution against a person or people on the basis of their membership of a particular racial or ethnic group, typically one that is a minority or marginalized." However, such definitions are problematic and can gloss over the deep structural problems that underpin society. Grosfoguel (2016), for example, urges us to see racism as the "materiality of domination used by the world-system" for the zone of being and the zone of not-being, whereas Schmid (1996) points to the behavioral, motivational, and cognitive features of racism. Oluo (2019, p. 11) writes "race is a social construct—it has no bearing in science" and is also a socioeconomic problem that has "woven its way into every part of our lives. It has shaped our past and our

DOI: 10.1201/9781003206958-4

futures." With such an immense impact on people's lives, how do AI systems even begin to understand race and racism?

AI codes can be made or interpreted by machines in a way that excludes and the people making the codes can of course be biased. In this chapter, we will look in particular at such exclusive mechanisms, but with the knowledge of AIs only being products of how we humans create them. AI systems can become racist if they build on data or logic that reflect racist societal structures. As we will see in this chapter, racist outcomes of AI systems can have devastating consequences, even if they are unintended or happen seemingly at random. However, when we have knowledge of how and what leads to exclusion, inclusive measures can be taken. Technology might reproduce systemic racism, therefore we must also be aware of what societal structures produce racism in the first place. In this chapter, I will take a closer look at different ways technology has gone wrong when dealing with racial issues.

AI'S PHRENOLOGICAL TURN

Phrenology, a racist pseudoscientific relic of the past, argued that mental traits, such as intelligence, could be determined by precise skull measurements. This discredited theory, however, has echoes in the present, such as the company Faception that claims to use facial recognition to determine a person's IQ. The company has been accused of using machine learning to create your average "racist uncle," who classifies people into categories such as "academic researcher, poker player, or terrorist" (Snyder, 2018)—as it is built on the quite controversial idea that a person's face can determine their personality. With knowledge systematization being socially constructed, we don't have to go back far to find examples of phrenological ideologies used to justify "scientific policies" of, for example, wanting to create apartheid states where the majority, or people in power, were held in higher consideration than "the other."

There have been several examples of AI providing quite racist outcomes, like the now well-known case of Google's AI imagine recognition software labeling Black people as "gorillas" Yahoo's Flickr service, which also auto-tagged Black people as "ape" and the Dachau Nazi concentration camp as a "jungle gym" (Hern, 2018), and facial recognition systems in passport control that often tell Asian people to "stop closing their eyes" (Howard & Borenstein, 2018). (Google's solution at the time, to simply remove the category of "gorilla" from its analysis was hardly adequate.) There have also been several reports on how AI "beauty filters" are changing people's skin to a whiter tone and adding more Caucasian-looking facial features (Wang, 2020). Another failed example is from the think tank diversity.ai, which had a goal of "preventing racial, age, gender, disability and other discrimination by humans and A.I. using the latest advances in Artificial Intelligence" by (1) "using AI to detect, analyze, prevent and combat human biases and discrimination" and (2) "using AI to prevent discrimination by AI." The think tank came under scrutiny after its 2016 international beauty contest where AI algorithms, given the mandate to select the most "objectively" beautiful human faces from over 6000 entrants from over 100 countries, returned with 82% white winners (Howard & Borenstein, 2018; Levin, 2016). This "cycle of colorism" is harmful, especially when perpetrating to young girls and women around the world that whiter is more beautiful, where "an ancient form of prejudice about skin color is flourishing in the modern internet age" (Wong, 2021), illustrated in Figure 4.1.

But is this a turn, or a return, to racist historical practices? Benjamin (2019a, p. 421) argues that health data fed into AI systems has a historical racial bias, for example, showing how in earlier times "the intention to deepen racial inequities was more explicit, today coded inequity is perpetuated precisely because those who design and adopt such tools are not thinking carefully about systemic racism" (Buck et al., 2019).

Mirror, mirror on the wall,
who's the FAIREST of them all?

Figure 4.1 When AI judges beauty.

FACIAL RECOGNITION BIAS

One of the most well-known examples of facial recognition bias comes from Joy Buolamwini's MIT project Gender Shades. Buolamwini investigated why her own face, with a dark complexion, was not detected until she put on a white mask, and wanted to see how well facial recognition technology identified people from different regions of the world to assess if there was a difference in how people of color were assessed (Buolamwini, 2017). Her study chose 1270 images as a benchmark for a gender classification performance test, with participation from 3 African and 3 Northern European countries, and labeled the faces into female and male categories and lighter and darker skin color categories, finding according to the gendershades.org website that:

> The subjects were then grouped by gender, skin type, and the intersection of gender and skin type [...] Three companies—IBM, Microsoft, and Face++—that offer gender classification

products were chosen for this evaluation. All companies perform better on males than females with an 8.1% to 20.6% difference in error rates. All companies perform better on lighter subjects as a whole than on darker subjects as a whole with an 11.8% to 19.2% difference in error rates.

http://gendershades.org/overview.html

The Gender Shades study shows a bias in the datasets that the system uses. When these datasets are predominately white males, then white males will be better classified. Buolamwini is also the founder of the "algorithmic justice league" that aims to fight algorithmic bias. There have also been multiple reports on, for example, webcams not being able to detect people with dark skin tones (CNN.com, 2009) and hand-soap dispensers not working for dark skin (Fussell, 2017). These are all examples of a larger systemic issue that Williams (2020, p. 574) describes as "systems like facial recognition, predictive policing, and biometrics are predicated on myriad human prejudicial biases and assumptions which must be named and interrogated prior to any innovation" where lineages of surveillance are based on the injustice of controlling specific groups of people, for example, based on the color of their skin.

BIAS IN LAW ENFORCEMENT SYSTEMS

Law enforcement agencies have had a long relationship with technologies of "whodunnit," using systems for finding out who the culprit of crimes might be—which, according to Williams (2020), has historic roots in racist models of policing. From fingerprints to lie detectors, the justice criminal system has many technologies used to investigate and prosecute. Few technologies are unbiased or without issues although, as is the case with gang databases, which are "rife with many problems such as dirty data, racial bias, interminable collateral consequences, and counterproductive outcomes" (Richardson, 2021). This was also the case for the COMPAS, a judicial sentencing system that was shown by the investigative journalism organization ProPublica to flag Black people as more likely to commit crimes than white people (Angwin et al., 2016). They give the example of an 18-year-old Black woman, Borden, who picked up a children's bike and rode it in suburbia Florida, and a 41-year-old white man, Prater, who had a previous conviction for armed robbery, who was caught shoplifting. Both Prater and Borden's crimes were charged for theft of $80 items, but Borden was rated as high risk for future crime, whereas Borden was rated low, as the COMPAS system predicted that Black defendants were 77% more likely to be at high risk for committing violent crimes and 45% more likely to commit crimes in general (Angwin et al., 2016; Broussard, 2018, p. 155). But, if women, in general, commit less crime than men, the system assigns greater predictive power to race rather than gender.

Being wrongly classified as a criminal can have wide impacts on one's life—it can lower one's credit scores, affect their offers of insurance or premiums, and make it harder to access employment:

> This means that a gang database designation can have a perpetual blacklist effect on individuals thus leading to differential treatment by private and public actors and inhibiting or completely foreclosing housing, educational, employment, financial, immigration, public benefits, and social opportunities for a significant period of time, if not indefinitely.
>
> (Richardson, 2021, p. 51)

See also Leyton's (2003) book chapter on "The New Blacklists: The Threat to Civil Liberties Posed by Gang Databases."

A more innocent example is from the 2016 Pokémon Go craze, where millions were out and about, walking and catching augmented reality Pokémon creatures who were hiding here and there in the real world, but could only be seen by looking through one's phone, heavily using AI systems to create user experiences across the globe. One problem was, however, that many "pokéstops" were located in people's backyards and private property, and players of color reported feeling at great risk:

> ...countless Black Men who have had the police called on them because they looked "suspicious" or wondering what a second amendment exercising individual might do if I walked past their window a 3rd or 4th time in search of a Jigglypuff. When my brain started combining the complexity of being Black in America with the real-world proposal of wandering and exploration that is designed into the gameplay of Pokemon GO, there was only one conclusion. I might die if I keep playing.
>
> (Akil, 2016)

Even with an innocent game like Pokémon Go, one can see how the sociocultural and the technical are connected. The game, with its technology translating the real world into an augmented reality, where strange Pokémon beings would pop up, thus adding an extra layer to the real world, also implies that societal norms and values were challenged (e.g., where BIPOC people could have freedom of movement in search for the Jigglypuffs). For some players of the game, it was not feasible to be out catching Pokémon late at night, as being Black and wandering across people's lawns could in the US be deemed "suspicious."

AI IS NOT COLOR BLIND

Racism is not a single issue with quick policy fixes. Racism runs deep in society, and only by actively dealing with systematic, past, and present racist structuring of society can we have any hope that AI does not reproduce or invent new forms of racist work. In *Race After Technology*, Benjamin (2019b, p. backside) pinpoints how racism still defines American society within what she calls the "New Jim Code" of technology enabling racism, where technology such as AI is created within the racialized societal contexts, and "when (mis)used, worsen inequities for already marginalized people" showing how "technology reinforces systemic oppression in America, creating a digital dragnet which codes people by stigmatizing them for where they live, work, and play. Technology codifies discriminatory practices in a way that results in racist responses to social problems." Thus, we need to be extremely careful when implementing technology that is labeled as "color-blind" (i.e., "not focusing on race")—which is just sweeping deep societal issues under the rug.

Consider Obermeyer et al.'s (2019, p. 421) analysis of an AI cost of care prediction tool as a proxy for health needs. Such tools estimate how much will be spent on healthcare for an individual and use those costs as a way to estimate who has the greatest health needs. However, as Benjamin (2019a, p. 447) explains, such tools can have life-threatening results as "Black patients with the same risk score as White patients tend to be much sicker, because providers spend much less on their care overall." Thus, the tool may predict that less money will be spent on a Black person than a white person and thus conclude that the Black person is in better health, when in reality this is a symptom of both structural and interpersonal racism present in a healthcare system that is more inclined to spend resources on white patients. The Black person in this scenario may indeed have more health needs than the white person, thus using such a tool only compounds health inequalities.

Another example, from search algorithms, is Nobl's (2013, p. 1) study on the sexualization and colonial issues with how:

> Google Search results on the words 'Black girls' discursively reflect hegemonic social power and racist and sexist bias [...] which prioritize the interests of its commercial partners and advertisers, rather than rendering the social, political, and economic interests of Black women, and girls visible.

This is further described in her insightful book *Algorithms of Oppression* (Noble, 2018). A good impact of this research is that the search terms for Black women and girls are, at least for the present time, not sexualized in image searches—however, we need to go no further than to search for Asian women and girls to see that this is still a symptomatic issue even in the year 2022. Who is present, represented, and invited to the table is of uttermost importance (see e.g., Prescod-Weinstein [2020]'s work on white empiricism in physics, and the racialization of epistemology that it brings), and when AI learns from humans, we need to clean up our own species racism first.

SUMMARY

AI systems can have racist outcomes and be interpreted in racist ways, as technical codes can be written or interpreted by machines in a way that excludes certain groups of people. Whether it is willful programming actions to write programs trying to racialize intelligence based on faces or the unwilling silent discrimination of not having diverse input, such as the Gender Shades study—discrimination can be the outcome of programs. We have seen how law enforcement agencies are utilizing AI for mapping and assessing potential criminals, but in ways that can be discriminatory and assumptive. Seemingly innocent technologies, such as Pokémon Go, can also be exclusive due to social structures of racism. We must understand these systems in a wider context than as just something technical—as AI systems can become racist if they build on data or logic that reflects racist societal structures. One important first step is to ensure that racist data structures are not reproduced. A much larger endeavor is an actual societal change that limits the problem of structural racism in the first place. Technology can reproduce societal bias when issues like racism is not engaged as a structural problem.

REFERENCES

Akil, O. (2016). *Warning: Pokemon GO is a death sentence if you are a Black man*. https://medium.com/dayone-a-new-perspective/warning-pokemon-go-is-a-death-sentence-if-you-are-a-black-man-acacb4bdae7f

Angwin, J., Larson, J., Mattu, S., & Kirchner, L. (2016). *Machine bias: There's software used across the country to predict future criminals. And it's biased against Blacks.* Retrieved December 6, 2021 from https://www.propublica.org/article/machine-bias-risk-assessments-in-criminal-sentencing

Appiah, K. A. (2010). *The ethics of identity.* Princeton University Press.

Benjamin, R. (2019a). Assessing risk, automating racism. *Science, 366*(6464), 421–422.

Benjamin, R. (2019b). *Race after technology: Abolitionist tools for the new jim code.* Polity

Broussard, M. (2018). *Artificial unintelligence: How computers misunderstand the world.* MIT Press.

Buck, B., Scherer, E., Brian, R., Wang, R., Wang, W., Campbell, A., Choudhury, T., Hauser, M., Kane, J. M., & Ben-Zeev, D. (2019). Relationships between smartphone social behavior and relapse in schizophrenia: a preliminary report. *Schizophrenia Research, 208,* 167–172.

Buolamwini, J. A. (2017). *Gender shades: intersectional phenotypic and demographic evaluation of face datasets and gender classifiers.* MIT Press.

CNN.com. (2009). *HP looking into claim webcams can't see black people.* Retrieved December 06, 2021, from http://edition.cnn.com/2009/TECH/12/22/hp.webcams/index.html

Fussell, S. (2017). *Why can't this soap dispenser identify dark skin.* gizmodo. https://gizmodo.com/why-cant-this-soap-dispenser-identify-dark-skin-1797931773

Griffin, R. A. (2012). I am an angry Black woman: Black feminist autoethnography, voice, and resistance. *Women's Studies in Communication, 35*(2), 138–157.

Grosfoguel, R. (2016). What is racism? *Journal of World-Systems Research, 22*(1), 9–15.

Hern, A. (2018). Google's solution to accidental algorithmic racism: ban gorillas. *The Guardian.* Retrieved February 15, 2022 from https://www.theguardian.com/technology/2018/jan/12/google-racism-ban-gorilla-black-people

Howard, A., & Borenstein, J. (2018). The ugly truth about ourselves and our robot creations: the problem of bias and social inequity. *Science and Engineering Ethics, 24*(5), 1521–1536.

Levin, S. (2016). A beauty contest was judged by AI and the robots didn't like dark skin. *The Guardian,* 11(08). https://www.theguardian.com/technology/2016/sep/08/artificial-intelligence-beauty-contest-doesnt-like-black-people

Leyton, S. (2003). The new blacklists: The threat to civil liberties posed by gang databases. In Darnell F. Hawkins, Samuel L. Myers, Jr., & Randolph N. Stone (Eds.), *Crime control and social justice: The delicate balance* (pp. 109–174). Greenwood Press.

Miles, R. (2004). *Racism.* Routledge.

Noble, S. U. (2013). Google search: hyper-visibility as a means of rendering black women and girls invisible. *In Visible Culture,* 19.

Noble, S. U. (2018). *Algorithms of oppression.* New York University Press.

Obermeyer, Z., Powers, B., Vogeli, C., & Mullainathan, S. (2019). Dissecting racial bias in an algorithm used to manage the health of populations. *Science*, 366(6464), 447–453.

Oluo, I. (2019). *So you want to talk about race*. Hachette.

Prescod-Weinstein, C. (2020). Making Black women scientists under white empiricism: the racialization of epistemology in physics. *Signs: Journal of Women in Culture and Society*, 45(2), 421–447.

Richardson, R. (2021). Defining and demystifying automated decision systems. *Maryland Law Review*, 81, 785–840.

Schmid, W. T. (1996). The definition of racism. *Journal of Applied Philosophy*, 13(1), 31–40.

Snyder, B. (2018). https://twitter.com/jbensnyder/status/1064759174574280704

Wang, C. (2020). *Why do beauty filters make you look whiter?* Retrieved February 26 from https://www.popsci.com/story/technology/photo-filters-white-kodak-film/

West, C. (2017). *Race matters, 25th anniversary: With a new introduction*. Beacon Press.

Williams, D. P. (2020). Fitting the description: Historical and sociotechnical elements of facial recognition and anti-black surveillance. *Journal of Responsible Innovation*, 7(Supp1), 74–83.

Wong, J. (2021). *How digital beauty filters perpetuate colorism*. Retrieved March 1, 2022 from https://www.technologyreview.com/2021/08/15/1031804/digital-beauty-filters-photoshop-photo-editing-colorism-racism/

5

BODIES AND AI

HEALTH, AGING, AND DISABILITIES

In the dystopic *Matrix* movies, AI and machines have gone to war with mankind and won—harvesting our bodies as an energy source and trapping the humans that are left in virtual reality (VR). Luckily, that future imaginary of general AI as malignant and evil is far from the current narrow AI that we have today. However, the AI of the contemporary world still impacts our bodies, and in this chapter, I turn to how health, aging, and disability are impacted by AI systems.

How does growing old with AI work, who does it work for, and who does not benefit? And how does AI not only affect people's bodies growing old but also people with different bodily experiences? We need, for example, to open up the discourse to see, recognize, and discuss how people with disabilities are also impacted by AI. Bodily experiences are of course also important for younger generations, thus this chapter prompts a holistic view on health and how it is impacted by AI, but I will begin with the older age groups who are most often discussed in terms of AI and health, after getting a glimpse into the vast realm of medical AI and current debates.

DOI: 10.1201/9781003206958-5

ARTIFICIAL HEALTH OR INTELLIGENT HEALTH?

Healthcare informed by AI has the potential for novel advancements in diagnosis and treatment, but also risks overpromising results. It is also important to look at AI within the long history of discovery and innovation. Medicine is one of humanity's oldest sciences featuring a continuously evolving body of knowledge. Medicine's advancement has involved increasing practitioners' intelligence about how the body responds to disease and treatment. Thus, the evolution of healthcare has required an evolution of knowledge.

When it comes to healthcare, however, there is a major conundrum at work when AI is added to the mixture. Benjamin (2019, p. 421), for example, describes how "most hospital systems now utilize predictive tools to decide how to invest resources." Consider IBM's Watson, the Artificial Intelligence Division of Technology Behemoth IBM (named after the AI software Watson that won the TV quiz show Jeopardy in 2011), which is seeking to transform modern medicine using AI through different means. According to Strickland (2019), they have, however, "encountered a fundamental mismatch between the way machines learn and the way doctors work" and call the Watson Health story "a cautionary tale of hubris and hype"—as it was a prestigious project that overpromised but never achieved the results it needed to contribute to solving the healthcare issues it set out to. It is also criticized by Wachter (2015) in his book *Hope, Hype, and Harm at the Dawn of Medicine's Computer Age* to have a marketing-first, product-second approach. The meager results have led IBM to move away from AI diagnosis, according to Jay Royyuru, IBM's Vice President of Health Care and Life Sciences Research, it's "something the experts do pretty well. It's a hard task, and no matter how well you do it with AI, it's not going to displace the expert practitioner." (Strickland, 2019). After recommending "unsafe and incorrect' cancer treatments" (Ross & Ike, 2018), the way forward seems uncertain.

But AI can surprise. Take the AI system *BakedScan* as an example. Anyone who has entered a bakery in Japan is struck by a number of open-air displayed baked goods present. This is difficult for cashiers because they need to keep track of several different baked goods only by appearance. Enter BakedScan, a product developed by Hisashi Kambe, a Japanese, who developed BakedScan as an image recognition software to help cashiers identify which baked goods are where, with high accuracy, illustrated in Figure 5.1. Later, Kambe was contacted by a doctor who thought that cancer cells seen under a microscope looked suspiciously close to bread, and collaboration was funded using the same core system which identifies cancer based on images surprisingly well (Sommers, 2021). Another positive example is the software platform QUANTX which uses AI to identify breast cancer lesions (Jiang et al., 2021).

Strickland (2019) argues that a major issue with AI engaging in guesswork is that its "recommendations wouldn't be considered evidence based [...] Without the strict controls of a scientific study, such a finding would be considered only correlation, not causation." Other uses of AI in the medical field include robotic surgery for clinical-decision support, genetic analysis, medical administration,

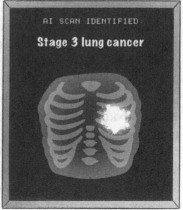

Figure 5.1 Bread and cancer.

and pathology for mental health issues, as well as personalized healthcare plans and faster hospital visits, AI robots revolutionizing endoscopy, and AI cloud-based digital drug discovery (Data-Core Healthcare, 2022). But before looking into concrete examples, I will first discuss how older adults as actual users of medical AI technology can be defined.

(RE)DEFINING THE 'ELDERLY'

These technologies for older adults are often termed *gerontechnologies* or *Assistive Technologies* (AT), which is any technology used to help people with disabilities or older adults. Other common terminologies are *Active Assistive Living* in the European Union or *welfare technology* in Scandinavia. AT encompasses a lot of different technologies from a cane to assist with walking, to hearing devices, glasses, and other sensory equipment, to more advanced robotics systems like social robots and exoskeletons. Some of these technologies have some sort of AI system built into them, and these are the ones that will be focused upon here.

Age is one of the easiest parameters to implement in computer systems and algorithms, as it most often has a common denominator: when a person was born and what date it is today (with some sociocultural exceptions, for example, Korean age, which counts as 1 year in the womb, and increases for everyone at New Year's). But in most cases, age is seemingly a simple matter. However, before we can quickly plot the age parameter into a computer system, let us unwrap the terminology. Who is actually "old" in today's societies? It differs both from individual to individual and also within societies. The countries with the longest life expectancies on the planet, for example, Japan (84.3 years on average) and Switzerland (83.7 years), are in stark contrast to the life expectancy of, for example, Sub-Saharan Africa on the lower end of the global spectrum with 23 year shorter lives on average (61.6 years) (Our World in Data, 2019). There's also a historic increase in life expectancy, but not much until the later part of the 19th century—up until then, life expectancy on all continents averaged around 25–35 years (although infant mortality of around 50% skews the data).

So who are "the elderly"—and is that even a good term to use? Östlund et al. (2015) argue that the term is derogatory and normative, and suggests "older adults" instead, a term with more agency and fewer other issues than "elderly." As we've seen demographically, life expectancy and medium age fluctuate, and setting a distinct threshold is difficult. When you think of the "elderly," certain prejudices might pop up in your mind. In many Western societies, the standard age of retirement is around 65–67 years of age, and it keeps increasing. However, this is far from universal, as members of some professions, such as politicians or university professors, often work well into their 80s, whereas other professions, such as professional dancers or law enforcement officers, retire younger due to bodily stress. However, whether accurate or not, the popular imagination often sees retirement as a major point of transition, with "retiree" being associated with "senior citizen" and being older. However, retiring from work does not mean retiring from life, as many Western countries are increasingly expecting that older adults will continue to be active well into older age, even in the case of illness that would have previously proved to be debilitating and necessitate institutionalization or severely restricting one's activities. Gerontechnologies can help facilitate that.

SENSORY MONITORING

One example of healthcare technology using AI is sensory monitoring. Habitual monitoring and upkeep are, for example, starting to become more widespread in dementia care. People with dementia[1] benefit from habitualization to keep cognitive functions running as long as possible, for example, by keeping routines and daily practices a common ground structure in their lives. This is, for example, why it's so difficult to move houses for a person with dementia, who quickly forgets new happenings and frames but retains memories going further back in time for longer. You can think of it as a retractable measuring tape, imagining your life as 1 cm equaling 1 year. If you were to get dementia on your 100th birthday, you would slowly retract memories, the 90–100 cm (or years in this metaphor) disappears first, then the 80–90 cm, so great-grandchildren are forgotten before grandchildren, then children, but the phone number from your friends from elementary school can remain clear as day for much longer. Of course, there are varieties in this simplified example, and cognitive decline is a complex field of study (Deary et al., 2009; Salthouse, 2009; Williams et al., 2010).

How can AI and computer systems help alleviate difficulties in remembering? One technological innovation is reminding technology, where calendars or habitual synchronization work in harmony to remind the older adult with dementia to eat breakfast, take a walk at noon, or call their family to tell them how they are feeling. An example of this is robotic-based sensory systems, which can monitor habits, giving notifications to dependents and healthcare providers, see for eample, my study of what happened when a robot shaped like a plant was connected to a system checking whether or not fridge doors aware being opened, and other sensory data (Søraa et al., 2021). This is but one of many studies of how technology is being tested out to see how lives can improve, especially when it comes to aging in one's own home instead of being institutionalized before it is strictly necessary. AI holds great potential within the healthcare sector as a useful set of tools that can improve people's lives, for example, helping older adults live in their own homes longer before needing to be institutionalized—which for society at large is often reduced to an economic issue, but for the individual having to move goes much, much deeper. It's about moving away from home as a safety zone, losing contact with friends and family, and a major shift toward the end of one's life.

Technologies such as sensory systems for detection hold the potential to increase well-being—not just delay death. AI systems built into existing technologies that work without us noticing are perhaps some of the most efficient ones. The average phone user checks their phone about 120 times a day, and many with facial recognition unlock mechanisms. What if these 120 potential facial scans could also scan for acute illnesses? That's something the Hong Kong developed app "Fatal Recognition" from 2019 sought to help with, for example, by checking if one side of your face droops down, a common sign of someone having a stroke, which will then "alert you to call emergency services and contact your preset person of choice" (Yu et al., 2020). Privacy concerns and consent of the user are important for such applications—on the one hand, we all probably want to be healed as quickly as possible, but people are weary of signing off consent on where and how their face data is being used.

CARING FROM A DISTANCE

As the global population, especially in Western countries, but also East Asian countries, is aging rapidly, the prospect of people living healthy in their own homes as long as possible has become a big political priority. Institutionalizing is an expensive change of life. At an institution, healthcare services are centralized, whereas, for the older adult living at home, healthcare services must come to them. But it isn't just clinical healthcare—with blood samples, medicines, wound treatment, etc.,—that is part of care. The social interactional aspect of care, being able to see and interact with family, friends, and communities, is also an important part of especially mental care.

Care from a technological distance can also be through machines. A more visibly hands-on—or in-patient technology of applied AI is the American *da Vinci Surgical System* made in the year 2000, where a human operator doctor controls the robot's grips and arms to give precise operational procedures. Here, doctors and machines are working in tandem to give the best results. In its 21 years of operation, the system has received some criticism for being too expensive for many institutions, and it has received some lawsuits after people have died during surgery, prompting some ethical questions of regulatory law (Bourla et al., 2008; DiMaio et al., 2011; Maeso et al., 2010). Another issue in terms of diversity is who has access to this rather expensive machinery. In Norway, for example, the first da Vinci system was donated by Norwegian billionaires when one of them got prostate cancer and needed it.

MENTAL HEALTH ISSUES

Health, is for many, synonymous with physical health, but mental health is also important. Studies suggest that 20% of US adults live with a mental illness, with estimates of 6% more with undiagnosed mental illness (NAMI, 2020). European studies suggest that 1 in 6 Europeans report mental illness. Many East Asian countries have also been shown to have high suicide rates (e.g., South Korea with 28.6 suicides per 100,000 people, making it the 4th highest rate in the world [World-Population-Review, 2022]). However, there is a staggering deprioritization of mental health compared to physical health. Can AI help? Several studies would suggest so, from anxiety detection (Shen & Rudzicz, 2017), to suicide risk assessment (Just et al., 2017; Morales et al., 2019), and assessing people's depression through social media practices (De Choudhury et al., 2013; Moreno et al., 2011; Schwartz et al., 2014). D'Alfonso (2020) explains how

> data-driven AI methods can be employed to develop prediction/detection models for mental health conditions. In particular, an individual's 'digital exhaust', the data gathered from their numerous personal digital device and social media interactions, can be mined for behavioral or mental health insights.

He focuses on three main ways when AI is being developed for mental health: (1) personal sensing or digital phenotyping, (2) natural language processing of clinical texts and social media content, and (3) chatbots.

(1) Personal sensing or digital phenotyping (the observable physical properties of people through digital means), which D'Alfonso (2020) describes as: "to infer contextual and behavioral information about an individual that can then be used as input for machine learning methods to predict psychological/psychometric outcomes and mental health conditions." These pieces of information can be retrieved, for example, by smartwatches and other tools that monitor sleep patterns (Difrancesco et al., 2019; Scott et al., 2019), or smartphone typing patterns for determining depression (Mastoras et al., 2019). Another study, by Buck et al. (2019), suggests that reductions in the number and length of outgoing phone calls could be associated with relapses of schizophrenia. One step further according to D'Alfonso (2020), "Ecological Momentary Interventions (EMI) is to provide momentary psychological interventions or behavioral prompts delivered via personal mobile devices during an individual's daily life, often informed by their responses."

(2) Natural language processing of clinical texts and social media content, where analyses of how (and what) a person is speaking about something can potentially help deduce their mental state (Calvo et al., 2017). One place to gather such information can be through social media, as D'Alfonso (2020) describes as: "By analyzing linguistic features in social media content, it is possible to generate machine learning models that can be used to infer an individual's mental health earlier than traditional approaches" (see also Briand et al., 2018; Eichstaedt et al., 2018; Ziwei & Chua, 2019).

(3) Chatbots have also been an area where mental health issues have been discussed in relation to AI, as the history of chatbots is tied to psychology, with the first chatbot ELIZA made to simulate a Rogerian psychotherapist (Weizenbaum, 1966). Scoping reviews suggest good results (Abd-Alrazaq et al., 2019), and with several chatbots developed for mental health purposes such as Replika, Tess, Woebot, and Wysa, users are in the millions, however, ethical issues are pronging (Fiske et al., 2019; Kretzschmar et al., 2019). For all AI-based mental health work, D'Alfonso (2020) warns that "If not kept in check, AI could exacerbate traditional ethical problems in mental health care."

Chen et al. (2019), for example, found in their article called "Can AI Help Reduce Disparities in General Medical and Mental Health Care?" that health care improves for some but might not improve for all, and warns against bias in the data and modeling choices. In regard to diversity among users, it can be beneficial to ask "which users" and how they are involved. If up to a quarter of all people suffer from a mental illness—some studies suggest that half of the US population will develop a mental illness during their lifetime (Reeves et al., 2011)—this is an unprecedented health crisis to be taken with the uttermost seriousness. However, mental health is still stigmatized among certain population groups. It is far less likely for men (globally) to seek help and talk about mental health issues (Rickwood et al., 2007; Vogel et al., 2014). One such problematization of how "mental health problems of racial groups often depend heavily on the larger social context in which the group is embedded" is seen by Chen et al. (2019), for example, whereas intersectional issues persist with young BIPOC men (Black, Indigenous, and People of Color) being the least likely to seek mental help (Rickwood et al., 2007).

When developing chatbots, or other natural language processing AI tools, awareness of forgotten user groups is key. Particularly when it comes to socioeconomically disadvantaged men—an often-overlooked group in diversity and exclusion contexts—much work remains to be done. It is, however, troubling in the diversity and AI context that groups who might benefit most from help are difficult to reach and to include as users—partly due to toxic masculinity with beliefs of mental troubles being effeminized in public discourse.

COVID, AI, AND HEALTHCARE DIVERSITY

The COVID-19 pandemic has globally been the most significant and impactful happening since the World Wars. With millions dead and hundreds of millions infected, the effects have been devastating on global health, economy, and future outlook. Community loss has been devastating, with higher losses from older adults and marginalized ethnic groups. Amidst these terrible losses, AI has played both the roles of hero and villain during the situation. On the one hand, AI has enabled quick detection of infection, helping to identify and quickly alert health authorities when new outbreaks were emerging—as can be seen in the case of South Korean uses of smartphone technology to map movement patterns and positioning. Robots have been actively employed for cleaning and disinfecting surfaces, for example, in Danish hospitals, and also to clean streets and deliver goods, such as food deliveries as seen in China and the US. Google's COVID-19 Open Data Repository, "a comprehensive, open-source resource of COVID-19 epidemiological data and related variables like economic indicators or population statistics from over 50 countries," is but one of many examples of Big Data powered AI possibilities for learning from this pandemic to prepare for future ones (Chou, 2022).

On the other hand, AI has also been used in the guise of pandemic safety to increase surveillance and monitoring. Some countries have used regulations against the spread of fake news to arrest protestors—both those protesting against mask use and those protesting against oppressive government policies. The COVID-19 pandemic has made it clear that we are in desperate need of AI regulations if we are to utilize the enormous possibilities the technologies bring, particularly to ensure that inclusion and diversity measures are taken. Technological solutions are not enough; consider, for example, how Sweden, one of the richest and most educated countries on the planet, effectively "failed the scientific method" and made unjust choices, which resulted in death rates ten times higher than neighboring Norway from the pandemic (Brusselaers et al., 2022).

Writing about ongoing events like this is a fine balance—we do not yet know how societies are reacting and reshaping to "the new normal" of living in a pandemic society, however, in terms of AI usage, it showcases really well how technology can be used for both good and bad. In terms of diversity, the pandemic does, of course, discriminate, in what we have described as a geronticide of deprioritizing older adults in the pandemic discourse (Søraa et al., 2020), where the older members of populations' life values are diminished due to socioeconomic gain and pain from lockdowns. However, it is not only the old who are badly impacted by the pandemic. Young people are particularly vulnerable and many of them have now had their entire high school or university experience as a digital experience, which is overly described as negative to their mental health (Creswell et al., 2021). Excluded also from the rich nations' hoarding of medicine is the Global South, which has waited for the Global North to be vaccinated one, two, or even three times before they got access to vaccines. This shows how health is not fair, how healthcare is biased, and how any technology implemented in the real world faces major diversity and exclusion issues from the get-go.

AI AND DISABILITIES

Care technologies are not only for older adults but also for another major user group, i.e., people with disabilities, regardless of their age. Disability is defined as "a physical or mental condition that limits a person's movements, senses, or activities." (by the Oxford's Lexico Dictionary). This can be physical like lack of walking mobility, lacking the sense of smell, or mental, like inability to understand social cues. Whittaker et al. (2019, pp. 2–3) in their thorough paper on "Disability, Bias, and AI" provides several key questions to include when discussing this intersection, which shortened is:

(1) How can we draw on disability activism and scholarship to ensure that we protect people who fall outside of the "norms" reflected and constructed by AI systems? (2) [and] recognizing that the intersections of race, gender, sexual orientation, and class often mediate how "disability" is defined and understood? (3) What standards of "normal" and "ability" are produced and enforced by specific AI systems? (4) How can we better highlight [...] consequences of being diagnosed and pathologized [with] opportunities to "opt out" before? (5) how might other anti-bias legal frameworks [..,] provide us with new ways to imagine and fight for accountability in AI, in service of disabled

people? (6) What tactics can those working toward AI account-ability learn from the disability rights movement (7) What kinds of design, research, and AI engineering practices could produce more desirable futures for disabled people?"

This highly practical guide can help reflect and chart pathways for better disability inclusion and diversity awareness. See also Hamraie and Fritsch (2019), who discuss the need for critique and reinvention of our material-discursive world to take into account justice for people with disabilities when constructing technology. People with disabilities face a wide range of discrimination daily—and with AI, the dangers are that these exclusive mechanisms are reproduced, or that new exclusive spaces or systems are created. Consider the lack of wheelchair ramps across the built environment, and how difficult it can be to get around in cities where walking on stairs become a barrier that prohibits immobile people to move freely about. Or, consider a work conference where wine tasting is the only social event, excluding people with a lack of taste or smell (as well as people who don't drink). Or on a wider scale, consider the educational system built around social interaction with classmates.

For people with disabilities, we can really see how applied AI can be utilized to improve lives. To illustrate, let us consider the case of people with a lack of sight. This can be as mundane as color-blind people (8% of men) struggling when websites use colors to differentiate things, which a color-blind person might not be aware of. Whether totally or slightly impaired, there are numerous ways blind or partly blind people can benefit from AI. Voiced reading of the text is one example, such as "Seeing AI," Microsoft's solution for helping blind people to gain information about their surroundings through their smartphone, which has shown (Granquist et al., 2021) Akiko Ishii, a blind Japanese mother of a young child, for example, uses the app to pick out clothes for her daughter by scanning for colors; she can get a live description of what her daughter is doing through video (e.g., "reading a book, playing with toys") and can benefit from the app helping to read the text to enable bedtime readings with her daughter (Tezuka, 2020). Another example is how sign language is being improved by AI capabilities (Jiang et al., 2020; Michaud et al., 2000; Parton, 2006), for example, by making AI that interprets sign language hand gestures into spoken words for conversation partners who don't understand sign language.

For neurodivergent individuals, AI also holds great potential, for example, for clinical decision support (Pedersen et al., 2020), for helping assess and helping users with autism learn to better interact with other people (Al Banna et al., 2020; Anagnostopoulou et al., 2020; Wall et al., 2012), but we also need to be aware of ethical blind spots when "automating autism" (Keyes, 2020)—translating social realities into computational models can lose aspects along the way. Likewise, people with mobility disabilities who have major benefits from being able to partly walk again from exoskeleton technology should also be critically explored to ensure that the technology works for all, for example, also users who are deemed too heavy or tall by the currently available technology (Lajeunesse et al., 2016; Søraa & Fosch-Villaronga, 2020). To reflect on how the world is configured in a way that presumes that everyone has bodies that work "perfectly" think about webpages, and how so many are made in ways where people with vision impairments or color-blindness might struggle to get accurate information

The body is facing several AI systems—how we age, how we are assumed to be able-bodied, and how our health is being quantified by technology. And, will such aging in and through AI be exclusive to the rich, securing eternal existence only for the one with the resources to pay for it? As we will see in the next chapter, AI discriminates already in life on how rich users are—and exclude nonusers based on access to capital. Also in health: "many AI applications that have been deployed in high-income country contexts, use in resource-poor settings remains relatively nascent" (Wahl et al., 2018). However, there are also examples of AI primarily targeting lower-income countries and individuals, for example, GraceHealth, a fertility app using AI smart technology and science-based predictions (Grace.health, n.d.).

Disability shapes our world—especially how it is discussed, governed, and excluded from power. The discrimination toward disabled people—ableism—"shapes our digital and technological imaginations - notions of who will 'benefit' from the development of [AI, which are] designed and implemented [...] how we envision the 'proper' functioning of bodies and minds" (Shew, 2020, p. 41). This techno-ableism is a close relative of tech-bias, which we have seen permutates sociotechnical society.

DEATH WITH AI—HEAVEN IS A PLACE ON EARTH?

Lastly, in this chapter, I'd like to take you on a thought experiment through the Black Mirror episode, San Junipero (Season 3, Episode 4). This British TV show explores the pitfalls of technologies—what happens when you stare into a black screen and see your own shortcomings and fears through a "black mirror" reflected from the screen? Most episodes are quite gloomy, but ironically, the episode that most strongly focuses on death and dying is quite high-spirited and uplifting. The episode explores youthful, introverted Yorkie falling in love with and exploring a beautiful 80s coastal town with the stunning and bubbly Kelly. The twist is, however, that both these young women are quite old, and dying, connected to the joyful party world of San Junipero through an AI-powered VR system. The episode has sparked several scholarly debates (Cook, 2019; Drage, 2018; Galati, 2021) – and is illustrated in Figure 5.2.

The big topic the episode portrays is eternal, digital life. Given the choice to wither and die, and to upload your conscience to live on forever in a "happy place"—much in parallel to the concept of "heaven" or "nirvana." As with most Black Mirror conundrums, fiction is closer to reality than we might think. In Japan, the so-called "hikikomori" i.e. young people with severe social withdrawal, along with other socially reclusive people in other countries, are escaping to virtual worlds of the internet for gaming and net socializing in Massive Multiplayer Online Roleplaying Games (MMORPGS), foregoing the digital world. Some millionaires are turning to cryogenic

Figure 5.2 San Junipero's take on death and AI.

freeing of their bodies to try to cheat death and be defrosted if/when technology advances to a point where it would be possible for them to wake up and live eternally. AI fuels both cryogenics, MMORPGS, and digital eternity portrayed in San Junipero. The social, and even death, is digitalized.

Fictional depictions of caring in connection to AI leave us with several questions: who is choosing what type of eternity one goes to? Can AI systems choose the wrong scenarios, for example, thinking that you love dogs but secretly you are terrified of them, leaving you in an eternal dog hell? Can you pull the plug if you get tired of digital existence?

SUMMARY

In this chapter, we have seen how bodies are treated by AI systems, focusing on health, age, and disabilities as three major topics where our bodies move in and with different systems. We saw how older adults can use gerontechnology to better their lives. Additionally, the impacts of the COVID-19 pandemic and its AI impacts were introduced, and we saw that health is a big marketplace when it comes to AI, but that there are still multiple difficulties ahead for implementing health AI that works well for all. For people with disabilities, we saw how AI can be quite beneficial, for example, when helping deaf or blind people communicate, but that AI often preassumes able-bodied individuals, and that awareness of people with disabilities is needed. Disabled people's lived experience is important but often discounted. However, this is a missed opportunity to engage impacted people as experts of their own lives and lived experiences. Drawing on Harawayan thinking, whereas many AI projects seek to do good, clearly listening to, and working with, those impacted by it would make any-*body*'s bodies *matter* in the chimera of our sociotechnical cat's cradle.

NOTE

1   Dementia is not a specific disease but is rather a general term for the impaired ability to remember, think, or make decisions that interferes with doing everyday activities. Alzheimer's disease is the most common type of dementia. Though dementia mostly affects older adults, it is not a part of normal aging. (CDC. (2019).

REFERENCES

Abd-Alrazaq, A. A., Alajlani, M., Alalwan, A. A., Bewick, B. M., Gardner, P., & Househ, M. (2019). An overview of the features of chatbots in mental health: A scoping review. *International Journal of Medical Informatics*, 132, 103978.

Al Banna, M., Ghosh, T., Taher, K. A., Kaiser, M. S., & Mahmud, M. (2020). A monitoring system for patients of autism spectrum disorder using artificial intelligence. *International Conference on Brain Informatics*. 19 September 2020, Padua, Italy.

Anagnostopoulou, P., Alexandropoulou, V., Lorentzou, G., Lykothanasi, A., Ntaountaki, P., & Drigas, A. (2020). Artificial intelligence in autism assessment. *International Journal of Emerging Technologies in Learning (iJET)*, 15(6), 95–107.

Benjamin, R. (2019). Assessing risk, automating racism. *Science*, 366(6464), 421–422.

Bourla, D. H., Hubschman, J. P., Culjat, M., Tsirbas, A., Gupta, A., & Schwartz, S. D. (2008). Feasibility study of intraocular robotic surgery with the da Vinci surgical system. *Retina*, 28(1), 154–158.

Briand, A., Almeida, H., & Meurs, M.-J. (2018). Analysis of social media posts for early detection of mental health conditions. *The 31st Canadian Conference on Artificial Intelligence took place in Toronto*. 8–11 May 2018, York University, Ontario.

Brüsselaers, N., Steadson, D., Bjorklund, K., Breland, S., Stilhoff Sörensen, J., Ewing, A., Bergmann, S., & Steineck, G. (2022). Evaluation of science advice during the COVID-19 pandemic in Sweden. *Humanities and Social Sciences Communications*, 9(1), 1–17.

Buck, B., Scherer, E., Brian, R., Wang, R., Wang, W., Campbell, A., Choudhury, T., Hauser, M., Kane, J. M., & Ben-Zeev, D. (2019). Relationships between smartphone social behavior and relapse in schizophrenia: A preliminary report. *Schizophrenia Research*, 208, 167–172.

Calvo, R. A., Milne, D. N., Hussain, M. S., & Christensen, H. (2017). Natural language processing in mental health applications using non-clinical texts. *Natural Language Engineering*, 23(5), 649–685.

CDC. (2019). *What is dementia?* Centers for Disease Control and Prevention. Retrieved May 28, 2022 from https://www.cdc.gov/aging/dementia/index. html#:~:text=Dementia%20is%20not%20a%20specific,a%20part%20 of%20normal%20aging.

Chen, I. Y., Szolovits, P., & Ghassemi, M. (2019). Can AI help reduce disparities in general medical and mental health care? *AMA Journal of Ethics*, 21(2), 167–179.

Chou, K. (2022). *Making data useful for public health*. Retrieved March 19, 2022 from https://blog.google/technology/health/making-data-useful-public-health/

Cook, J. (2019). San Junipero and the digital afterlife: Could heaven be a place on Earth? In David Kyle Johnson (Ed.), *Black Mirror and Philosophy: Dark Reflections*, 109–117. John Wiley & Sons, Inc.

Creswell, C., Shum, A., Pearcey, S., Skripkauskaite, S., Patalay, P., & Waite, P. (2021). Young people's mental health during the COVID-19 pandemic. *The Lancet Child & Adolescent Health*, 5(8), 535–537.

D'Alfonso, S. (2020). AI in mental health. *Current Opinion in Psychology*, 36, 112–117.

Data-Core-Healthcare. (2022). *Success stories of using artificial intelligence in healthcare*. Retrieved May 28, 2022 from https://datacorehealthcare.com/success-stories-of-using-artificial-intelligence-in-healthcare/

De Choudhury, M., Counts, S., & Horvitz, E. (2013). Predicting postpartum changes in emotion and behavior via social media. *Proceedings of the SIGCHI conference on human factors in computing systems*. 27 April–2 May 2013, Paris, France.

Deary, I. J., Corley, J., Gow, A. J., Harris, S. E., Houlihan, L. M., Marioni, R. E., Penke, L., Rafnsson, S. B., & Starr, J. M. (2009). Age-associated cognitive decline. *British Medical Bulletin*, 92(1), 135–152.

Difrancesco, S., Lamers, F., Riese, H., Merikangas, K. R., Beekman, A. T., van Hemert, A. M., Schoevers, R. A., & Penninx, B. W. (2019). Sleep, circadian rhythm, and physical activity patterns in depressive and anxiety disorders: A 2-week ambulatory assessment study. *Depression and Anxiety*, 36(10), 975–986.

DiMaio, S., Hanuschik, M., & Kreaden, U. (2011). The da Vinci surgical system. In Jacob Rosen, Blake Hannaford, & Richard M. Satava (Eds.), *Surgical robotics* (pp. 199–217). Springer.

Drage, E. (2018). Utopia, race, and gender in Black Mirror's "San Junipero". In Barry Vacker, & Angela M. Cirucci (Eds.), *Black mirror and critical media theory*, (p. 27). Lexington Books.

Eichstaedt, J. C., Smith, R. J., Merchant, R. M., Ungar, L. H., Crutchley, P., Preoţiuc-Pietro, D., Asch, D. A., & Schwartz, H. A. (2018). Facebook language predicts depression in medical records. *Proceedings of the National Academy of Sciences*, 115(44), 11203–11208.

Fiske, A., Henningsen, P., & Buyx, A. (2019). Your robot therapist will see you now: ethical implications of embodied artificial intelligence in psychiatry, psychology, and psychotherapy. *Journal of Medical Internet Research*, 21(5), e13216.

Galati, G. (2021). San Junipero. In German A. Duarte, & Justin Michael Battin (Eds.), *Reading Black Mirror* (pp. 291–308). Transcript-Verlag.

Grace.health. n.d. Reshaping health for the next billion women online. *The first women's health platform designed for emerging markets*. https://www.grace.health/

Granquist, C., Sun, S. Y., Montezuma, S. R., Tran, T. M., Gage, R., & Legge, G. E. (2021). Evaluation and comparison of artificial intelligence vision aids: Orcam MyEye 1 and seeing AI. *Journal of Visual Impairment & Blindness*, 115(4), 277–285.

Hamraie, A., & Fritsch, K. (2019). Crip technoscience manifesto. *Catalyst: Feminism, Theory, Technoscience*, 5(1), 1–33.

Jiang, X., Satapathy, S. C., Yang, L., Wang, S.-H., & Zhang, Y.-D. (2020). A survey on artificial intelligence in Chinese sign language recognition. *Arabian Journal for Science and Engineering*, 45(12), 9859–9894.

Jiang, Y., Edwards, A. V., & Newstead, G. M. (2021). Artificial intelligence applied to breast MRI for improved diagnosis. *Radiology*, 298(1), 38–46.

Just, M. A., Pan, L., Cherkassky, V. L., McMakin, D. L., Cha, C., Nock, M. K., & Brent, D. (2017). Machine learning of neural representations of suicide and emotion concepts identifies suicidal youth. *Nature Human Behaviour*, 1(12), 911–919.

Keyes, O. (2020). Automating autism: Disability, discourse, and artificial intelligence. *The Journal of Sociotechnical Critique*, 1(1), 8.

Kretzschmar, K., Tyroll, H., Pavarini, G., Manzini, A., Singh, I., & Group, N. Y. P. S. A. (2019). Can your phone be your therapist? Young people's ethical perspectives on the use of fully automated conversational agents (chatbots) in mental health support. *Biomedical Informatics Insights*, 11, 1178222619829083.

Lajeunesse, V., Vincent, C., Routhier, F., Careau, E., & Michaud, F. (2016). Exoskeletons' design and usefulness evidence according to a systematic review of lower limb exoskeletons used for functional mobility by people with spinal cord injury. *Disability and Rehabilitation: Assistive Technology*, 11(7), 535–547.

Maeso, S., Reza, M., Mayol, J. A., Blasco, J. A., Guerra, M., Andradas, E., & Plana, M. N. (2010). Efficacy of the Da Vinci surgical system in abdominal surgery compared with that of laparoscopy: a systematic review and meta-analysis. *Annals of Surgery*, 252(2), 254–262.

Mastoras, R.-E., Iakovakis, D., Hadjidimitriou, S., Charisis, V., Kassie, S., Alsaadi, T., Khandoker, A., & Hadjileontiadis, L. J. (2019). Touchscreen typing pattern analysis for remote detection of the depressive tendency. *Scientific Reports*, 9(1), 1–12.

Michaud, L. N., McCoy, K. F., & Pennington, C. A. (2000). An intelligent tutoring system for deaf learners of written English. *The 4th ACM SIGCAPH Conference on Assistive Technologies on assistive technologies*. 13–15 November 2000, Arlington Virginia USA.

Morales, M., Dey, P., Theisen, T., Belitz, D., & Chernova, N. (2019). An investigation of deep learning systems for suicide risk assessment. *Proceedings of the sixth workshop on computational linguistics and clinical psychology*. Minneapolis, Minnesota.

Moreno, M. A., Jelenchick, L. A., Egan, K. G., Cox, E., Young, H., Gannon, K. E., & Becker, T. (2011). Feeling bad on Facebook: Depression disclosures by college students on a social networking site. *Depression and Anxiety*, 28(6), 447–455.

NAMI. (2020). *Mental health by the numbers*. Retrieved May 28, 2022 from https://www.nami.org/mhstats

Östlund, B., Olander, E., Jonsson, O., & Frennert, S. (2015). STS-inspired design to meet the challenges of modern aging. Welfare technology as a tool to promote user driven innovations or another way to keep older users hostage? *Technological Forecasting and Social Change*, 93, 82–90.

Our-world-in-data. (2019). *Life expectancy, 2019*. Retrieved March 19, 2022 from https://ourworldindata.org/grapher/life-expectancy-at-birth-total-years

Parton, B. S. (2006). Sign language recognition and translation: A multidisciplined approach from the field of artificial intelligence. *Journal of Deaf Studies and Deaf Education*, 11(1), 94–101.

Pedersen, M., Verspoor, K., Jenkinson, M., Law, M., Abbott, D. F., & Jackson, G. D. (2020). Artificial intelligence for clinical decision support in neurology. *Brain Communications*, 2(2), fcaa096.

Reeves, W. C., Pratt, L. A., Thompson, W., Ahluwalia, I. B., Dhingra, S. S., McKnight-Eily, L. R., Harrison, L., D'Angelo, D. V., Williams, L., & Morrow, B. (2011). Mental illness surveillance among adults in the United States. *MMWR. Morbidity and Mortality Weekly Report*. https://stacks.cdc.gov/view/cdc/31627

Rickwood, D. J., Deane, F. P., & Wilson, C. J. (2007). When and how do young people seek professional help for mental health problems? *Medical Journal of Australia*, 187(S7), S35–S39.

Ross, C., & Ike, S. (2018). IBM's Watson supercomputer recommended 'unsafe and incorrect' cancer treatments, internal documents show. Statnews.com. Retrieved January 27, 2022 from https://www.statnews.com/2018/07/25/ibm-watson-recommended-unsafe-incorrect-treatments/

Salthouse, T. A. (2009). When does age-related cognitive decline begin? Neurobiology of Aging, 30(4), 507–514.

Schwartz, H. A., Eichstaedt, J., Kern, M., Park, G., Sap, M., Stillwell, D., Kosinski, M., & Ungar, L. (2014). Towards assessing changes in degree of depression through Facebook. Proceedings of the workshop on computational linguistics and clinical psychology: From linguistic signal to clinical reality (pp. 118–125). June, Baltimore, Maryland, USA.

Scott, J., Grierson, A., Gehue, L., Kallestad, H., MacMillan, I., & Hickie, I. (2019). Can consumer grade activity devices replace research grade acti-watches in youth mental health settings? Sleep and Biological Rhythms, 17(2), 223–232.

Shen, J. H., & Rudzicz, F. (2017). Detecting anxiety through reddit. Proceedings of the fourth workshop on computational linguistics and clinical psychology—From linguistic signal to clinical reality.

Shew, A. (2020). Ableism, technoableism, and future AI. IEEE Technology and Society Magazine, 39(1), 40–85.

Sommers, J. (2021). The pastry A.I. that learned to fight Cancer. Retrieved January 25, 2022 from https://www.newyorker.com/tech/annals-of-technology/the-pastry-ai-that-learned-to-fight-cancer

Søraa, R. A., & Fosch-Villaronga, E. (2020). Exoskeletons for all: The interplay between exoskeletons, inclusion, gender, and intersectionality. Paladyn, Journal of Behavioral Robotics, 11(1), 217–227.

Søraa, R. A., Manzi, F., Kharas, M. W., Marchetti, A., Massaro, D., Riva, G., & Serrano, J. A. (2020). Othering and deprioritizing older adults' lives: Ageist discourses during the COVID-19 pandemic. Europe's Journal of Psychology, 16(4), 532.

Søraa, R. A., Nyvoll, P., Tøndel, G., Fosch-Villaronga, E., & Serrano, A. (2021). The social dimension of domesticating technology: Interactions between older adults, caregivers, and robots in the home [Vitenskapelig artikkel]. Technological Forecasting and Social Change, 167, 1–13. https://doi.org/10.1016/j.techfore.2021.120678

Strickland, E. (2019). IBM Watson, heal thyself: How IBM overpromised and underdelivered on AI health care. IEEE Spectrum, 56(4), 24–31.

Tezuka, K. (2020). Seeing AI empowers people who are blind or with low vision for everyday life. Retrieved April 16, 2022 from https://news.microsoft.com/apac/2020/12/03/seeing-ai-empowers-people-who-are-blind-or-with-low-vision-for-everyday-life/#:~:text=Power%20of%20artificial%20intelligence, nearby%20people%2C%20text%20and%20objects.

Vogel, D. L., Wester, S. R., Hammer, J. H., & Downing-Matibag, T. M. (2014). Referring men to seek help: The influence of gender role conflict and stigma. *Psychology of Men & Masculinity*, 15(1), 60.

Wachter, R. (2015). The digital doctor. *Hope, Hype and at the Dawn of Medicines Computer Age*, 2015.

Wahl, B., Cossy-Gantner, A., Germann, S., & Schwalbe, N. R. (2018). Artificial intelligence (AI) and global health: how can AI contribute to health in resource-poor settings? *BMJ Global Health*, 3(4), e000798.

Wall, D. P., Dally, R., Luyster, R., Jung, J.-Y., & DeLuca, T. F. (2012). Use of artificial intelligence to shorten the behavioral diagnosis of autism. *PLoS ONE*, 7, e43855.

Weizenbaum, J. (1966). ELIZA—A computer program for the study of natural language communication between man and machine. *Communications of the ACM*, 9(1), 36–45.

Whittaker, M., Alper, M., Bennett, C. L., Hendren, S., Kaziunas, L., Mills, M., Morris, M. R., Rankin, J., Rogers, E., & Salas, M. (2019). *Disability, bias, and AI*. AI Now Institute. https://www.ahrq.gov/downloads/pub/evidence/pdf/alzheimers/alzcog.pdf

Williams, J. W., Plassman, B. L., Burke, J., & Benjamin, S. (2010). Preventing Alzheimer's disease and cognitive decline. *Evidence Report/Technology Assessment*, 193, 1–727.

World-Population-Review. (2022). *Suicide rate by country 2022*. Retrieved May 28 2022 from https://worldpopulationreview.com/country-rankings/suicide-rate-by-country

Yu, M., Cai, T., Huang, X., Wong, K., Volpi, J., Wang, J. Z., & Wong, S. T. (2020). Toward rapid stroke diagnosis with multimodal deep learning. *International Conference on Medical Image Computing and Computer-Assisted Intervention*. 4–8 October 2020, Lima, Peru.

Ziwei, B. Y., & Chua, H. N. (2019). An application for classifying depression in tweets. *Proceedings of the 2nd international conference on computing and big data*. 18–20 October 2019, Taiwan.

6

AI AND CLASS

WORK, EDUCATION, AND SUSTAINABILITY

Resources are unequally dispersed among societal groups. AI risks benefitting the rich and disadvantaged or poorer members of the population. At the same time, the AI system depends on raw materials to be extracted and the logistics of work that poorer groups are often set to do (Crawford, 2021). While the rich get access to a choice of technologies, the poor are more likely to have the technology implemented in their lives without having a say in what form or quantity the implementation takes. This is especially risky for AI systems dealing with surveillance, where inequality can make the rich the surveyor through technology and the poor the surveyed. This chapter treats "class" as a sociocultural phenomenon (that consists of a myriad of aspects, including the intersection with gender, race, age, etc.) but what I will focus primarily on is the socioeconomic aspect of the class, both within individual societies and globally. Wealth, poverty, and income division are some of the biggest exclusion mechanisms on the planet. Let us then consider what pays off and who is kept in debt when AI enters the economic part of being human.

DOI: 10.1201/9781003206958-6

COMPUTER SAYS NO-MONEY, NO-GO

AI systems are good at finding and reproducing patterns and structures and sorting the world through those frames, but less good at making ethical choices (if at all possible, which is a topic philosophers struggle with). Given a demographic set of 100 people, if 10 are super rich and 50 are poor, that is just data and information, but how the algorithms further deal with this data is where the trouble begins. Let us consider an example of AI systems being deployed to figure out who has the right to a mortgage or insurance where the only criteria are that individuals should be above certain thresholds of income levels and their programming does not allow them to do a holistic assessment of the individuals. AIs can reinforce poverty, leading to poorer insurance for the poor and a debt spiral of always having to rent because of not being granted a mortgage, or even becoming homeless as a result of not getting access to that mortgage. Being homeless also has a whole set of discriminatory sociotechnical discrimination, as Rosenberger (2017) describes as "callous objects: designs against the homeless" (e.g., hostile urban environments through benches one cannot sleep on).

What can be done? One example of intervention in the system can be seen in Norwegian banks, where state-imposed limits of 15% self-equity of a property price before it can hand out mortgage loans. However, the banks are allowed to deviate from this strict rule in 10% of the cases, and decisions can be made based, for example, on cases not fitting the strict system (e.g., high earners, who have not saved up enough yet, but who most likely will do so quickly). Whereas the system is strict with a clear high bar, the human factor allows a humane case worker to decide that the odds of a poor person being able to afford to pay a mortgage regardless of their saving value is high enough to grant them a mortgage. Some aspects of the economy are already highly automated. Consider taxation—a horde of human accountants might come to mind, peering over tax papers and cards, carefully calculating who owes what to whom—but this need not be the case. In Scandinavian countries, along with many others, taxes are mostly digital and algorithmically controlled.

The major problem with value-based assessment is that an individual's value is calculated based on their monetary value or, in some places, their social credit score, which measures how well they are performing as citizens based on whatever criteria a state decides is important. Some aspects of our lives, such as income, are more easily transferred to data variables than others. Some might be illegal to feed into computers, while some are given willingly, for example, images we upload through social media. For individuals not fitting the model of what the deciding powers deem valuable, exclusion from mortgages, loans, and benefits is at risk. Whether it is an AI system, a company, or state policy, if your value as a customer or citizen is devalued based on religion, sexuality, gender, race, or age, to give some examples, that is highly problematic and we need to take action to prevent AI from reproducing and inventing such bias. If AI "discovers" a negative correlation between people of a certain type and "bad" attributes, we can see how, for example, the system wasn't even intending to evaluate based on gender but ended up doing so. However, countermeasures do exist, such as Rikare 2, a Swedish project aiming to ensure gender-equal learning algorithms in financing decisions, or the AI platform "Alice" connecting female and minority entrepreneurs to help business upscaling (Ramboll, 2022, p. 15).

TOP-DOWN CONTROL IN THE WORKPLACE

AI is increasingly used in the employment sector to manage and control individual workers (Guenole & Feinzig, 2018; Kang et al., 2020). People (2020), for example, found that 24% of companies used AI for hiring purposes. This can, for example, be done through text analysis of recruitment applications (Dastin, 2018; Nimbekar et al., 2019), training of employees (Ku, 2021), and monitoring for unwanted worker behaviors (Eisenstadt, 2021). A study by van den Broek et al. (2021) of developers of AI for use in recruitment found that the technologists, although they were concerned with bias, treated it as a *human* phenomenon that "infects" the hiring process when decisions are made by humans based on subjective assumptions or emotional reactions. There have been ethnographic investigations of how employees experience technology and AI at work (Bhargava et al., 2021; Li et al., 2019; Bhargava et al., 2021; Li et al., 2019; Smith, 2018) and with developers of such technologies (Tomé et al., 2020; van den Broek et al., 2021), with most pointing in the direction of bias, if left unchecked, can have major consequences for disadvantaged people in power-asymmetric relations to their workplace.

Several large companies now use AI surveillance tools to monitor and "correct" the behavior of their employees. Take Amazon, for example, which is fighting unionization campaigns that have started partly as a way to challenge Amazon's micromanaging workers' time "on task," for example, tracking each second a worker spends on doing work. Bathroom breaks are excluded from the list of what constitutes "time on task," and so are mandatory meetings with supervisors (Delfanti, 2021; Kantor et al., 2021). This micromanagement of low-income workers is not new but follows a long line of Taylorist approaches made famous by an American behavioral engineer, Frederick Winslow Taylor, around the turn of the 20th century. Taylor, who was actively opposed to unionization, believed that workers were lazy by nature but, that through careful monitoring of every single movement in their workday and with proper incentives, they could be conditioned to work optimally (Littler, 1978).

Within warehouses managed by AI today, we find remnants of Taylorism on steroids—the technology allows for much more precise monitoring and ranking of employees than Taylor would have ever been able to imagine. This form of management shows that for a long time, human workers have been treated as part machines, entities that can be optimized, and as true machines have become increasingly optimized; there has been an expectation that the human worker follows suit. Workplace surveillance is not limited to physical work activities, as "employers can record their workers' every movement, listen in on their conversations, measure minute aspects of performance, and detect oppositional organizing activities" (Bales

& Stone, 2020, p. 1). Thus, the optimization we expect of workers, both derived from the way we control technology and aided by that very technology, can extend to every aspect of a worker's behavior. However, the mindset of continuously optimizing human work forgets that workers' lives are socially situated in ways that true machines are not (Delfanti, 2021; Gutelius & Theodore, 2019; Moradi & Levy, 2020).

Similarly to warehouses, transportation workers like cabbies, bus drivers, and truckers also face active monitoring of their workdays (Levy, 2016; Zhang et al., 2019). With the possibility of always knowing where the companies' vehicles are, the drivers of these vehicles are also closely monitored as they go where the vehicles go. This can have the beneficial aspect of ensuring that drivers take enough mandatory breaks, but it has also facilitated AI-driven driving assignments that are increasingly straining, inhuman, and not working well in human daily practices. But, let us not forget that AI-based logistic systems help plan and optimize people and goods moving through complex sociomaterial networks for our daily lives, sending packages from factories out, helping to avoid traffic jams with optimal route planning, and, thus, making sure your packages and you arrive on time.

Another example of unequal technology is autonomous cars that are often mentioned in the same breath as AI. Self-driving has been problematized, for example, by Crawford (2021), to still have a long way to go to be implemented fairly and safely into societies (not to mention, a lack of public debate if it is the wanted future). In terms of income diversity, access to different advanced transportation security systems is becoming important. AI systems that can emergency break by itself reroute if accidents happen ahead, or slow down when animals are spotted in the area, are in many ways reserved for the rich that can afford cars with these systems installed. The general issue of who AI cars should save has been problematized by MIT's moral machine experiment, finding, for example, that women are preferred to be saved over men, and that passengers are more valued than pedestrians. There is also a difference in cultures, for example, people in East Asian nations prefer to save older adults, whereas Europeans and Americans prefer that autonomous cars should save the young. If cars are to be autonomous, the question of who is responsible, and if the car should prioritize saving passengers or pedestrians is an ethical conundrum that continues to be discussed (Awad et al., 2018). Or, as Evans et al. (2020) suggest by seeing a self-driving car's decision-making as claim mitigation: "different road users hold different moral claims on the vehicle's behavior, and the vehicle must mitigate these claims as it makes decisions about its environment," where their goal is to "provide a computational approach that is flexible enough to accommodate several "moral positions" concerning what morality demands and what road users may expect." There are certainly ethical obligations for road safety, but also regarding access: when safety becomes a rich privilege, how does unsafe exclusion of poorer individuals impact society? And, how are societal values and ethics built into the technologies we create?

These cases of AI-aided top-down workplace control have a socioeconomic impact because many of the jobs that are most amenable to increased control are traditionally "low-skilled" jobs with low pay, whereas the individuals who design the technological systems of control and decide how they are implemented are highly paid. Thus, this is another example of people with high economic status controlling those with low economic status.

CLASS-BASED ACCESS TO EDUCATION

AI and computer systems hold great potential for knowledge development through the digital classroom. Massive Open Online Courses were already on the rise, but during the COVID-19 pandemic, they have had much increased relevance as the swap to digital classrooms made online teaching the new norm for many learners during lockdowns, quarantine, home offices, and the like. For some tutors, an issue can be that students never turn on their cameras during lectures, with the strange feeling of talking to black screens, whereas for low bandwidth areas, the opposite can be an issue if students don't turn off their cameras, leading to broadband connectivity issues. The internet is not a never-ending horn of abundance, and issues can come in many sociomaterial forms. O'Hara and Hall (2021) suggest that the internet can be conceptualized as four different internets that have rival values for governance and stability:

> Four Internets contends that the apparently monolithic 'Internet' is in fact maintained by four distinct value systems—the Silicon Valley Open Internet, the Brussels Bourgeois Internet, the DC Commercial Internet, and the Beijing Paternal Internet—competing to determine the future directions of internet affordances for freedom, innovation, security, and human rights.

Within this, it is easy to see that what we often think of as more or less established, static constants in digital reality, is changing and socioconstructed in negotiations between multiple actors.

In this technological scheme, it is important to understand access too, and the quality of education. Biased AIs can be present in the educational system, such as when the Boston Public Schools attempted to use ADS to combat racialized socioeconomic divides in their public school systems, but actually, it intensified segregation across the city's public schools (Aloisi & Gramano, 2019; Bales & Stone, 2020; De Stefano, 2019). Luckily, the project was evaluated to find the discrepancy; however, most public bodies that use ADS do not have the resources and/or prioritization for evaluation (Richardson, 2021). One possible solution or mitigation effort for better societal control of AI could be a mandatory evaluation of their effects, especially now in the technologies' infancy, where outcomes are not easy to predict.

It's not only students who can be impacted by AI in education, teacher evaluation systems have also been deployed to infer the quality of teachers based on student achievement in large-scale standardized tests (Richardson, 2021, p. 11). This grouping of teachers at specific schools or districts can create a false categorization of the effectiveness of education. This is problematic, as research shows that there is bias against female professors (Amrein-Beardsley, 2014; Richardson, 2021), as well as racial bias (Fisher et al., 2019; Kaschak, 1978). But if society is biased and the machine risks reproducing this bias (or indeed, creating new forms of bias unknown to humans), what can be done? That is a topic to be explored in the next chapter, but before that, let us consider other socioeconomic biases that AI can bring.

(UN)SUSTAINABLE AI

The discriminatory side of AI is not only a "sole matter" for the individual but also a "soil matter" for the sustainability of the socio-material world. In *Atlas of AI*, Fisher et al. (2019, p. 322) call for a holistic understanding of how AI is impacting global value chains of raw materials, for example, laborers extracting rare minerals to fuel lithium batteries, logistic networks of shipping these, for example, with underprivileged workers with few rights, and further chains of underpaid human workers in the loop of, for example, Amazon's Mechanical Turk program, with strict monitoring and surveillance work, for example, on how often one can use the restroom, lest it is retracted from salaries. Thus, Crawford and other holistic AI system investigators poignantly ask when is it enough. Are there ways to move forward without straining our ecosystems to continue producing more, better, faster, stronger computer systems and the hardware required for them? How can cycles of technological exploit, exemplified in Figure 6.1, be sustainable?

Technology is never neutral; it is often an accelerator of social differences. While the poor get poorer and the rich get richer, we must actively engage with AI as systemic oppression—asking who has access to technology, who is excluded from it, who is surveilled and controlled and by which entities—questioning the power they hold with a technological grasp over customers, inhabitants, or others. Not all inhabitants of the web are equal, and there are spiders crawling and gauging the poor denizens of the networks of the web, which is made possible by AI systems.

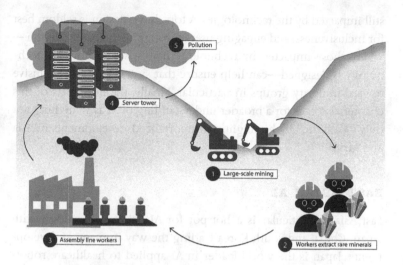

Figure 6.1 Socio-material cycle of AI.

There are also several ethical concerns, as Awad et al. (2018) and Nyholm and Smids (2016) warn in the book *AI & Ethics*. While AI technology is being developed, a question of importance is AI ownership and who benefits from it? Should a robot be taxed if taking over the job of a human for example? Where does the income of using robots flow, and which regulatory frames are put in place to ensure the dogma of "the rich getting richer, while the poor gets poorer"? Some scholars, politicians, and economists speak about universal basic income as the likely way to ensure that we do not get into a perfect storm of unequal resource allocations due to robotic revolutions and AI dominance.

Another major discrepancy in AI is the differences between public and private AI development, use, and regulation. While nation states and public companies might be to some extent more bound to their democratic (or autocratic) policies on how technology can be used on citizens, as well as noncitizens, for example, in the context of immigration, private companies also have a responsibility toward their customers, as well as those who are "non-users" but still impacted by the technologies. A top-down design is seldom best for inclusiveness, and engaging users through participatory design— where those impacted by technology have a say in how the technology is designed—can help ensure that systems are less exclusive toward minority groups in particular. Finally, the next parts of this chapter also calls for a broader understanding of AI as something not only "Western," by highlighting important AI-developments in Asia and Africa.

EAST ASIAN AI

East Asia in particular is a hotspot for AI-driven technology, with Japan, China, and South Korea leading the way in many AI developments. Japan is the world leader in AI applied to healthcare robotics in particular (Toshkov, 2018) (see also Chapter 5 on health in this book). China has taken leaps in a wide variety of infrastructural AI-centered systems, also now publishing more research papers than any other country on AI (27%) while at the same time filing for the most AI patents in the world. China specializes in speech and vision recognition, and with an AI-promoting policy backed by abundant funding opportunities, the country is putting its mark on the global AI field (Ema et al., 2016; Lundin & Eriksson, 2016; Okuda et al., 2013; Sato et al., 2020), however, with different privacy regulations than in Western countries (Li et al., 2019). As China is prepared to invest over $150 billion in AI projects in the next decade (Hao, 2019; Larson, 2018), it is clear that the future of AI will be greatly shaped by what the country decides to focus on.

South Korea has a particular focus on industrial AI-developed technology, i.e., automobiles, TVs, and smartphones, with Hyundai, LG, and Samsung leading the way. The country has a particular focus on soft power and entertainment (similarly to Japan's "Cool Japan" strategy) with, for example, the digital resurrection of deceased K-pop stars (Dräger & Müller-Eiselt, 2020), and, in particular, has utilized location technology innovatively with the COVID-19 pandemic (Choi, 2020).

AI IN AFRICA

Big Data and AI-based interventions are often thought of as innovations that emerge from the Global North and flow to the rest of the world. Coeckelbergh (2020) criticized this mindset in their call for "Parables of AI in/from the Global South." Terms like the "Global South" are useful in centering the fact that 75% of the world's population has access to only 20% of the global income and that most academic, political, and technological discourse does not include them. AI is of course not intrinsically a thing from the Global North neither in its making nor its usage; indeed digital technology holds a more borderless potential than its analog predecessors. But we risk reproducing such assumptions by not diversifying and situating studies, technology, and involved actors geographically to also include the Global South. Although many technologies and studies are developed and criticized by Westerners, there are a multitude of examples from the Global South. I acknowledge that the terms "Global South," "the West," and "developing countries" are all highly contested, colonial remnants, and they are not neutral terms. I will try as much as possible thus focusing on specific regions to be more precise.

"Global South" can mean many things at the same time, as Singh & Guzmán (2021) define it as:

[1] a site in its own right to study situated technological developments, and [2] a method to understand, analyze, and build developmental, postcolonial, and decolonial computing practices to (3) a metaphor for varying forms of suffering caused by capitalism and colonialism at a global scale, and [4] an effort to creatively resist and subvert such suffering.

For further readings on the terminology, read Singh and Guzmán (2021) and Toshkov's (2018) "The 'Global South' is a terrible term. Don't use it!"

Upon moving to Accra, Ghana, to start Africa's first Google AI lab, Moustapha Cisse describes how "lack of diversity can entrench unintended algorithmic biases and build discrimination into AI products." Cisse, a Senegalese trained in the USA, decided to cofound the organization Black in AI after experiencing that he was often one of the only Black people attending AI conferences in Europe drawing thousands of largely white attendees. This is another example of how structural biases, for example, visa denials to attend conferences abroad, work against African technologists. Cisse (2018) argues that keywords on AI from the Global North, such as bias and fairness, are seen as tools subject to human ingenuity. Priorities from the Global South, such as postcolonial computing (Singh and Guzmán, 2021; Ali, 2016) and decolonial computing (Irani et al., 2010) points to the importance of seeing computing as both situated and something that creates and maintains societal structure. At the same time, computing also holds the possibility for radical change, data extractivism, and seeing human dignity as the bridge between AI and human rights, data colonialism, and indigenous data sovereignty (Ali, 2016). Other important focus areas are feminist solidarity in design practices (Lovett et al., 2019), and data justice to focus on experiences with data-driven technologies as emerging ways of life. And as Kumar et al. (2019) argue, we need good data practices for indigenous data sovereignty and governance.

SUMMARY

This chapter has looked at socioeconomic issues of class that AI systems can perpetuate. AI can thus work as status cementing, as AI not programmed for it, will not yield new results but reproduce existing ones. AI systems used to assess if people should be given loans or mortgages will work with business as usual unless programmed to be allowed to take risks and individuals whose metrics do not measure up to the systems' parameters. We've also seen how workplaces are facing diversity issues due to AI top-down control, where workers with less power are further controlled not only by bosses and managers but also by AI systems that are given control over workers, especially where the socioeconomic potential for higher salaries is low. This class-based discrimination can also be seen in the education system, where plans for change can also have unintended consequences, showing the complexity of implanting AI for change. We also saw how AI is not only an individual issue but a sustainable issue for the whole planet, cementing class differences through soil extraction of resources used for the infrastructure AI relies on, and we saw how AI is a global phenomenon. Thus, AI can be a resource for the rich to get richer but be a serious barrier for poor people.

REFERENCES

Ali, S. M. (2016). A brief introduction to decolonial computing. *XRDS: Crossroads, The ACM Magazine for Students*, 22(4), 16–21.

Aloisi, A., & Gramano, E. (2019). Artificial intelligence is watching you at work: Digital surveillance, employee monitoring, and regulatory issues in the EU context. *Comp. Lab. L. & Pol'y J.*, 41, 95.

Amrein-Beardsley, A. (2014). *Rethinking value-added models in education: Critical perspectives on tests and assessment-based accountability.* Routledge.

Awad, E., Dsouza, S., Kim, R., Schulz, J., Henrich, J., Shariff, A., Bonnefon, J.-F., & Rahwan, I. (2018). The moral machine experiment. *Nature*, 563(7729), 59–64.

Bales, R. A., & Stone, K. V. (2020). The invisible web at work: Artificial intelligence and electronic surveillance in the workplace. *Berkeley Journal of. Employment. & Labour. Law*, 41, 1.

Bhargava, A., Bester, M., & Bolton, L. (2021). Employees' perceptions of the implementation of robotics, artificial intelligence, and automation (RAIA) on job satisfaction, job security, and employability. *Journal of Technology in Behavioral Science*, 6(1), 106–113.

Choi, J. (2020). Why Covid-19 only accelerates South Korea's AI ambitions. *The Interpreter*. Retrieved March 1, 2022 from https://www.lowyinstitute.org/the-interpreter/why-covid-19-only-accelerates-south-korea-s-ai-ambitions

Cisse, M. (2018). Look to Africa to advance artificial intelligence. *Nature*, 562(7728), 461–462.

Coeckelbergh, M. (2020). *AI ethics*. MIT Press.

Crawford, K. (2021). *The atlas of AI.* Yale University Press.

Dastin, J. (2018). *Amazon scraps secret AI recruiting tool that showed bias against women.* Reuters.com. Retrieved December 6, 2021 from https://www.reuters.com/article/us-amazon-com-jobs-automation-insight-idUSKCN1MK08G

De Stefano, V. (2019). "Negotiating the algorithm": Automation, artificial intelligence, and labor protection. *Comparative. Labour. Law. & Policy Journal*, 41, 15.

Delfanti, A. (2021). Machinic dispossession and augmented despotism: Digital work in an Amazon warehouse. *New Media & Society*, 23(1), 39–55.

Dräger, J., & Müller-Eiselt, R. (2020). *We humans and the intelligent machines: How algorithms shape our lives and how we can make good use of them*. Verlag Bertelsmann Stiftung.

Eisenstadt, L. F. (2021). # MeTooBots and the AI workplace. *University of Pennsylvania Journal of Business Law, Forthcoming*. Available at SSRN: https://ssrn.com/abstract=3921186

Ema, A., Akiya, N., Osawa, H., Hattori, H., Oie, S., Ichise, R., Kanzaki, N., Kukita, M., Saijo, R., & Takushi, O. (2016). Future relations between humans and artificial intelligence: A stakeholder opinion survey in Japan. *IEEE Technology and Society Magazine*, 35(4), 68–75.

Evans, K., de Moura, N., Chauvier, S., Chatila, R., & Dogan, E. (2020). Ethical decision making in autonomous vehicles: The AV ethics project. *Science and Engineering Ethics*, 26(6), 3285–3312.

Fisher, A. N., Stinson, D. A., & Kalajdzic, A. (2019). Unpacking backlash: Individual and contextual moderators of bias against female professors. *Basic and Applied Social Psychology*, 41(5), 305–325.

Guenole, N., & Feinzig, S. (2018). The business case for ai in HR. Retrieved November 4, 2020, from https://www.ibm.com/downloads/cas/AGKXJX6M.

Gutelius, B., & Theodore, N. (2019). The future of warehouse work: technological change in the US logistics industry. *UC Berkeley Labor Center*.

Hao, K. (2019). China has started a grand experiment in AI education. It could reshape how the world learns. *MIT Technology Review*, 123(1). https://www.technologyreview.com/2019/08/02/131198/china-squirrel-has-started-a-grand-experiment-in-ai-education-it-could-reshape-how-the/

Irani, L., Vertesi, J., Dourish, P., Philip, K., & Grinter, R. E. (2010). Postcolonial computing: a lens on design and development. *Proceedings of the SIGCHI conference on human factors in computing systems*. 10–15 April 2010, Atlanta Georgia USA.

Kang, Y., Cai, Z., Tan, C.-W., Huang, Q., & Liu, H. (2020). Natural language processing (NLP) in management research: A literature review. *Journal of Management Analytics*, 7(2), 139–172.

Kantor, J., Weise, K., & Ashford, G. (2021). The Amazon that customers don't see. *The New York Times*.

Kaschak, E. (1978). Sex bias in student evaluations of college professors. *Psychology of Women Quarterly*, 2(3), 235–243.

Ku, L. (2021). *An introduction of NLP and how it's changing the future of HR*. Plug and Play. https://www.plugandplaytechcenter.com/resources/introduction-nlp-and-how-its-changing-future-hr/

Kumar, N., Karusala, N., Ismail, A., Wong-Villacres, M., & Vishwanath, A. (2019). Engaging feminist solidarity for comparative research, design, and practice. *Proceedings of the ACM on Human-Computer Interaction*, 3(CSCW), 1–24.

Larson, C. (2018). China's AI imperative. *Science*, 359, 628–630.

Levy, K.E.C. (2016). Digital surveillance in the hypermasculine workplace. *Feminist Media Studies*, 16(2), 361–365.

Li, J. J., Bonn, M. A., & Ye, B. H. (2019). Hotel employee's artificial intelligence and robotics awareness and its impact on turnover intention: The moderating roles of perceived organizational support and competitive psychological climate. *Tourism Management*, 73, 172–181.

Littler, C. R. (1978). Understanding taylorism. *British Journal of Sociology*, 29, 185–202.

Lovett, R., Lee, V., Kukutai, T., Cormack, D., Rainie, S. C., & Walker, J. (2019). Good data practices for Indigenous data sovereignty and governance. In Angela Daly, Monique Mann, & S. Kate Devitt (Eds.), *Good Data*, 26–36.

Lundin, M., & Eriksson, S. (2016). Artificial intelligence in Japan (R&D, market and industry analysis). *EU-JAPAN Centre For Industrial Cooperation*, 39, 26–36.

Moradi, P., & Levy, K. (2020). *The future of work in the age of AI: Displacement or risk-Shifting?* Frank Pasquale, Sunit Das, Markus Dirk Dubber (Eds.). Oxford University Press.

Nimbekar, R., Patil, Y., Prabhu, R., & Mulla, S. (2019). Automated resume evaluation system using NLP. *2019 International Conference on Advances in Computing, Communication and Control (ICAC3)*. 20–21 December 2019, Mumbai, India.

Nyholm, S., & Smids, J. (2016). The ethics of accident-algorithms for self-driving cars: An applied trolley problem? *Ethical Theory and Moral Practice*, 19(5), 1275–1289.

O'Hara, K., & Hall, W. (2021). *Four internets: data, geopolitics, and the governance of cyberspace*. Oxford University Press.

Okuda, T., Shiotani, S., Sakamoto, N., & Kobayashi, T. (2013). Background and current status of postmortem imaging in Japan: short history of "Autopsy imaging (Ai)". *Forensic Science International*, 225(1-3), 3–8.

People, S. (2020). *The changing face of HR: HR and people leaders' report*. Sage.com. https://www.sage.com/en-gb/blog/the-changing-face-of-hr/

Ramboll. (2022). *AI for gender equality—Addressing inequality through AI*. https://www.vinnova.se/globalassets/mikrosajter/ai-for-jamstalldhet-starker-tillvaxten-samhallsekonomin-och-arbetsmarknaden/ramboll---ai-for-gender-equality-2020-11-19.pdf

Richardson, R. (2021). Defining and demystifying automated decision systems. *Maryland Law Review, Forthcoming*, 81, 785–840.

Rosenberger, R. (2017). *Callous objects: Designs against the homeless*. University of Minnesota Press.

Sato, M., Yasuhara, Y., Osaka, K., Ito, H., Dino, M. J. S., Ong, I. L., Zhao, Y., & Tanioka, T. (2020). Rehabilitation care with Pepper humanoid robot: A qualitative case study of older patients with schizophrenia and/or dementia in Japan. *Enfermeria Clinica*, 30, 32–36.

Singh, R., & Guzmán, R. L. (2021). *Parables of AI in/from the Global South*. Retrieved December 7, 2021 from https://www.4sonline.org/parables-of-ai-in-from-the-global-south/

Smith, C. (2018). An employee's best friend? How AI can boost employee engagement and performance. *Strategic HR Review*.

Tomé, E., Rivera, O., Lopez, D., & de Mello, P. S. (2020). Robotics and artificial intelligence (R&Ai) perceptions of consumers and producers: An international comparison among Portugal and Spain. *Proceedings of the European conference on the impact of artificial intelligence and robotics, ECIAIR*.

Toshkov, D. (2018). The 'global south' is a terrible term. Don't use it! *Research Design Matters*. http://re-design.dimiter.eu/?p=969

van den Broek, E., Sergeeva, A., & Huysman, M. (2021). When the machine meets the expert: An Ethnography of developing ai for hiring. *MIS Quarterly*, 45(3), 1557–1580.

Zhang, C., Lu, Y., Feng, M., & Wu, M. (2019). Trucker behavior security surveillance pbased on human parsing. *IEEE Access*, 7, 97526–97535.

7

INTERSECTIONALITY AND
RESPONSIBLE AI

Artificial Intelligence is not neutral but laden with sociocultural norms and values, some of which might create bias and discriminatory results. Previous chapters show that multiple issues of diversity, exclusion, and bias are at play when AI is implemented into complex societal settings. At the same time, the AI debate is filled with (human) emotion. On the one hand, you find the techno-optimism of Silicon Valley and programming cultures that forget the societal part of the sociotechnical connection. On the other hand, you find critical voices with little understanding of what AI means and what it can and cannot do, with Luditte's vendetta against any technological change. And then there are a myriad of voices and opinions in between. For such disruptive technologies as AI, we need knowledge from experts as well as laypeople. My aim with this book has been twofold: demystify a lot of what AI is and unwrap who is included or excluded in different AI systems, and which diversity issues arise. In this final chapter, I will take a step back to see the big picture and point the reader to different ways to analyze how AI and diversity impact your life, organization, or topic of research, and pathways that can take us toward more inclusive and responsible societies with AI.

DOI: 10.1201/9781003206958-7

INTERSECTIONAL ISSUES WITH AI

The users of technology will always find ways of using it differently than intended, and who is imagined to be the standard user often doesn't become the actual user in reality. In that regard, it is important to debunk the myth that a user is "just" one personal characteristic. Although this book is divided into chapters on the topics of gender, queernes, race, bodies, and class, it does not mean that these are indistinct characteristics that do not impact each other. Additionally, it is important to stress that such division should not facilitate essentialism.

We need to look at how these characteristics impact each other on structural, practice, and semantic levels. Intersectionality can be an important methodological and theoretical tool for this (Lovett et al., 2019; Crenshaw, 1990, Mahoney et al., 2019; May, 2015). Intersectionality can "reveal both the intersections of institutions, systems, and categorizations that produce oppression and the intersections of identity categorizations within individuals and groups" (Runyan, 2018). Intersectional issues can lead to double discrimination, for example, by management with inclusion programs that include Black men and white women, but not Black women. This intersectional discrimination is then difficult to argue against, if companies did include Black people and women. There are multiple examples of how intersectional exclusion can disadvantage people with two or more marginalized personal characteristics or identities, for example, disabled and older gay people (Runyan, 2018), women with AIDS (Dressel et al., 1997), and clashes of race, class, and gender (Margulies, 1994) (Figure 7.1).

Throughout this book, we have seen many examples of how a person is not just one category, but rather that it is important to treat personhood as complex and contextually dependent. My argument is not about the individual, but rather that it is important to see the sociotechnical structuring, discrimination, and marginalization that technology let loose can imply. Applying the

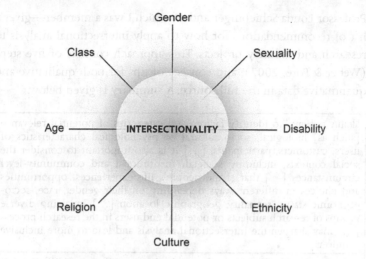

Figure 7.1 Intersectionality.

categorizations used in scholarly writing on gender, class, race, etc., is out of necessity, to mobilize our understandings of identities to provide discourse for structural change. Five years ago, noting that hand sanitizer sensors work worse for people of color (PoC) than for white people might have been an example of technological discrimination, but treated as minor. However, in the context of a global pandemic, the functionality of hand sanitizers suddenly became of high importance, and thus racial disparities could have a large impact. Likewise, hardware like tablets and phones that work worse for older adults with more wrinkles on both face and hands can also accelerate bad user experiences and create distrust in technologies. In another example we saw in the book, hand and face identifiers were not working well for PoC.

An intersectional approach can lead to more inclusive research and engineering solutions by better predicting user needs (Cuadraz & Uttal, 1999). The Stanford University and the European Commission of Research's expert panel on Gendered Innovations—led by

Professor Londa Schiebinger and of which I was a member—gives a list of recommendations for how to apply intersectional analysis to research and industry projects. This approach consists of five steps (Weber & Fore, 2007), and consists of tips for both qualitative and quantitative data in the full source. A summary is given below.

Identify problems "to identify factors and categories of potential relevance [such as] biological, socio-cultural or psychological characteristics of users, customers, participants. [...] It is also important to consider the social contexts, including societal, institutional and community-level circumstances [...] that shape people's life experiences, opportunities and choices in different ways depending on their gender, race, socio-economic status, sexuality, geographic location [...] Involving diverse groups of research subjects or potential end users in the research process [...] may sharpen the intersectional analysis and lead to more inclusive solutions."
Design research "to illuminate the multiplicative effects of different but interdependent categories and factors. Determine which methods [...] are best suited for examining the intersecting variables [...] factors may vary by social context, and may change over time. [Check] questions and categories for misguided or stereotypical assumptions before initiating the data collection."
Collect data "Applying an intersectional approach presupposes data collection on factors intersecting with sex and gender (e.g., ethnicity, religion, sexual behavior, lifestyle, socioeconomic status, disability, gender categories relevant to transgender people, etc.). The attention to individual-level categories and characteristics should be complemented by a focus on group-level factors (e.g., at the household, neighborhood, institutional, regional, state or national level [...] In qualitative research, the sample should be heterogeneous enough to capture the various intersecting positions of relevance to the research problem."
Analyze "to illuminate the multiplicative effects of different but interdependent categories and factors. Quantitative research should move beyond an additive focus on main effects (e.g., estimating separate effects for gender, race and sexual orientation) to examine how the variables in focus intersect [...]. Qualitative analyses will often be exploratory in nature and should offer rich descriptive accounts of the various categories, factors and processes that intersect to shape people's identities, opportunities and practices in a given context."
Disseminate "Reporting should specify [...] how information for each variable was obtained. To promote transparency, researchers should report all relevant outcomes of the intersectional analysis including inconclusive results."

This list is but one of many from the Gendered Innovations toolbox; for further investigation, see the website *genderedinnovations.stanford.edu*. In the particular interest of AI researchers and students, the case studied engineering on "Assistive technologies for the elderly, Extended virtual reality, Facial recognition, Gendering social robot, Machine learning, Machine translation, and Virtual assistants," to see case studies on analyzing gender and intersectionality.

With an intersectional approach to AI, there are a couple of things to pay especially close attention to: (1) Are there blind spots in the data or interpretation of the data due to "token-clusters" (e.g., having datasets of many PoC, many women, but few Black women)? Or, having datasets of many people from the LGBTQ+ community (if relevant), many older adults, but no older gay men? People can be counted into marginalized groups, but since people are much more diverse than just the color of their skin, their age, sexuality, gender, etc., we risk blind spots. (2) Is tokenism used to justify the inclusion of marginalized groups, or is there deeper respect and value for why we need diverse datasets in the first place? Data and digital tools are for people, not the other way around, thus having strong recommendations, guidelines, and in some cases, regulations that technology companies can steer toward can help bring about more inclusive AI. (3) Is intersectionality seen as a "one time fix" or a continuous action of work, where continuous checks, groundings, and interventions are needed to ensure that new forms of exclusion do not arise? All these three issues have been shown to be present in diversity issues relating to AI, and an intersectional approach can help mitigate some of the exclusion and discrimination issues that arise.

Intersectionality is not without controversy as a term. It is, for example, seen by some as too ambiguous and lacking clear outcomes. One study of the 2017 Women's March on Washington (Gendered Innovations, 2020) looked at how intersectionality was attempted to be used to create a political agenda for "all" women, whereas Vardeman and Sebesta (2020) urge us to problematize presuppositions in their analysis of how debates regarding intersectionality "enable self-reflexivity, positionality and critique, but also risk becoming routinized gestures in activist and academic." Making and implementing good frameworks for intersectionality is a challenging task, as it is often resisted, misunderstood, and misapplied, both by critics and supporters (Aldrin Salskov, 2020).

TOWARD SOCIOTECHNICAL DIVERSITY IN AI?

This book could have been titled many things. "AI & bias" or "AI & exclusion" would have put a focus on the negative sides of what AI could be but would, in some sense, put an overall negative policing of everything that is wrong with AI as a premise. On the other hand, calling the book "AI & inclusion" or "AI & social change" would have given a false predisposition of AI as benevolent, and those issues would have an easy fix. Instead, I wanted to tread a middle ground, to be aware of the bad sides of AI and how it can be biased but also chart possibilities. This Rogerian method of aiming to establish a middle ground between opposing viewpoints is rather difficult, as each side of the debate will necessarily disagree with some parts of the book, as no good deed goes unpunished when dealing with wicked problems—i.e., social problems that are so complex and interconnected that they become nearly impossible to solve. It's easy to lose ourselves in negativity. However, a problematizing approach that can take an informed, middle path toward diversity in AI is needed.

As a scholar from *Science and Technology Studies* (STS) and posthuman *Feminist Technoscience* thinking, there are four main points I would like to make. As we have seen throughout this book, the technologies we use in our everyday lives are shaped by (1) the people making them, which makes diverse representation behind the creations crucial, when analyzing a technology's *script* (Akrich et al., 2002; Fallan, 2008; Verbeek, 2005). Likewise, (2) the *user-perspective* and enrollment of co-development is important, to directly benefit the groups one designs for, but also adhering to deprioritized users, and marginalized groups, specifically when looking at how users *domesticate* technology (Berker et al., 2005; Søraa et al., 2021; Sørensen, 1994).

Thirdly, (3) we need an understanding of the sociomaterial network of actors, both human and nonhuman, who are part of constructing what AI means and how it is being used in societies.

Seeing (1) producers, (2) user perspectives, and (3) the wider networks in which the technology takes a path, some threads of diversity "pains and gains" start emerging, which can lead us to a deeper understanding of how AI technology is (4) situated. This situatedness is complex, as situated data is constructed, framed, and processed for different audiences and purposes (Rettberg, 2020) and situated knowledge is always made in a societal context (Haraway, 1988), which also draws on what technology could or ought to be. Sociotechnical diversity thus implies critical thinking as the cornerstone of analyzing how AI impacts different social groups, and how it can reproduce bias and exclusion, situated in specific contexts.

Diversity in AI can be implemented before, during, and post-processing of data. Looking back at the Gender Shades example where Buolamwini (2017) explains how facial recognition software discriminates both against women and PoC—and thus also in an intersectoral way, particularly for women of color—we can see how Western situated technology has built-in biases against people not categorized as the "standard user." Here, when the datasets implemented before processing are biased, the out-data also necessarily becomes biased if measurements are not taken to de-bias the data. We see here that this is a double-edged sword. If the user is configured with "I-methodology" (of white males imagining their users to also be white males), or if they try to imagine the user as "everybody," one still puts "somebody" over "anybody" (Buolamwini, 2017)—thus coding the male as "default" (Oudshoorn et al., 2004). Regardless, if the focus is on every-, some-, or any-"body"—bodies that matter (Perez, 2019) are still situated in societal contexts and are impacted by AI. At the same time, we must acknowledge that "diversity" is a contested term, with its own problems and criticisms.

As we departed from in Chapter 1, diversity can be defined as the condition of having or being composed of differing elements, especially the inclusion of different people in a group or organization. Therefore, using a diversity approach entails considering a wide variety of personal characteristics such as gender, sexual orientation, age, race/ethnicity, physical/mental abilities and health, socioeconomic background, and religion/spirituality, to mention some. But this "division" is not without issues, as it can reproduce sociocultural understandings of the world from biased humans, thus also creating biased AI. When data is fed into a computer program, labeling person X as "a queer Black immigrant man," the program itself does not necessarily have any deeper understanding of what those parameters *mean*, or how they are socially constructed. Being an immigrant, for example, requires some background knowledge: immigrating from where to where? Why is the person on the move? Is it due to famine or a lucrative job opportunity? We see time and again that AIs often produce racist results when the predisposition is to categorize, for example, immigrants as something negative. Likewise, an AI wouldn't really know what being "queer" here means, as understandings of gender in AI systems are often confused by "biological sex," which should be problematized when basic websites ask "are you a man or a woman." For gender-fluid or nonbinary people, this strict categorization of "gender = sex" just doesn't make sense. Likewise, for the parameter "immigrant," would an AI know when the person would stop being an immigrant—is it when they cross the border, or for example when they get new citizenship? What about their grandchildren, who might still be socially flagged as "different" when they are presented with racist questions like "but where are you *really* from?"

My point aligns with Crawford's (2021) statement that AI is neither artificial nor intelligent. Its sociomateriality is clear, as resources are extracted from the earth to construct machines, software, and systems that AI relies on to function, drawn out from the earth's minerals, and put together by human hands and labor. Neither is it intelligent if one judges based on what we understand as human intelligence. As we saw in Chapter 1, the definition that divides narrow AI as being used nowadays to solve specific problems, vs. general AI that can think and act like humans, which is something we don't know if it will ever exist. It is a sociotechnical imaginary in the making, with complex threads to fiction, futurist dreams, and visions, and it is enacted in laboratories and companies throughout the world as a dream of a synthetic being with intelligence. It has never been made before and might never be made in the future. When it does, we need to throw our analytical minds toward it, but before that, let us consider how actual (narrow) AI is impacting contemporary and near-future societies. As my book aims to show, diversity is one of the key impacts we need to scrutinize when it comes to AI.

But, as we saw in Chapter 1, diversity has been criticized as a term for not being inclusive enough (being invited to the party but not to dance), so whenever inclusivity is spoken of, we must be reflective of who and what one means, and in what ways including the specific group is important. Diversification and inclusion procedures need to be continually used, challenged, and reconceptualized, and with AI, there is no "quick fix" to magically make one's data, algorithms, and output "diversified once and for all." Society is constantly changing, old biases are uncovered and scrutinized, new values are made and prioritized, knowledge, data, actions, and imaginaries are situated, and if we can move away from this "solutionism" of machines as Crawford (2021) describes, to reconfigure our human relations to machines, then we are really on to something—but what steps can be taken for making AI more just and responsible for all?

PATHWAYS TOWARD RESPONSIBLE AI?

Before concluding this book, I will reflect on diversity responsibilities in AI, before providing you, the critical reader, with some further resources and matters to follow in your further knowledge journey about the topic of diversity and AI. Many AI books, especially in the social sciences and humanities academic disciplines, are quite pessimistic. Although, as indeed, I have shown throughout this book, there are major societal issues connected with AI systems when it is designed or used willingly or with unaware evil outcomes. But let us not forget that AI also has a positive impact on society, with health care diagnostics and personalized recommendations, for example, through AI system tools and robotic surgery. Navigational and logistic systems are daily rerouting, planning, and optimizing people and goods moving through complex networks, avoiding traffic jams and making sure both you and your packages arrive on time. AI systems for text editing make sure our communication stays crisp, and tools for ensuring good grammar are especially useful.

If you grew up before the 3rd millennium, you might remember the world without Google and search algorithms—where information was much more difficult to find and more time-consuming. Everyone has an opinion on social media, but a world without it is less social—at least in terms of outreach and expanding your core group of friends and family. Other examples include e-commerce, e-health, and e-learning. I do not write about AI critically because I have a big dislike for it. Quite the contrary, I believe that a critically constructive approach can help us improve and build better AI systems—to benefit humanity.

For AI to be responsible, we need all actors involved in its lifecycle to also be responsible. This implies that the developers, designers, producers, and makers of AI must carefully think through what they make, why, who it benefits in what ways, and who it excludes. It implies that users, both end-users, secondary users, random users, early adopters, latecomers, nonusers, and everyone who in some way or the other will use AI or be impacted by its use are able to voice their opinion about what this means for their lives and to be taken seriously. It implies that policymakers, governing forces, leaders, and administrators take seriously the impact AI has on systems that they manage or control. It impacts the ground workers who make AI's sociomaterial aspects, digs up minerals from the earth, dives into dangerous places to make AI run smoothly, and enables the neat and clean AI through dull, dirty, or dangerous work, which is the hidden side of making AI work. We are all humans and nonhumans in the sociomaterial network of AI, responsible for its responsibility. Responsible AI implies in some sense that AI could be responsible in itself. This is an exciting philosophical question that AI ethicists, philosophers, theologians, and the everyday person on the Internet have an opinion on. See for example Broussard (2018, p. 194) for a general overview of the robot rights debate, Gunkel (2018) for a study on AI and rights arguing for "robot rights now," Darling (2016) who argues for extending legal protection to social robots due to their effects of anthropomorphism, Gellers (2020) for an AI take on animals and environmental law, and Smith (2021, 2022) for a theological discourse on the topic. For pathways to responsible AI, I would argue that there are a couple of things to be aware of:

(1) We must be wary of *technological solutionism*, thinking that for any societal issue, there is a technological quick fix. Sometimes, to solve issues of diversity, exclusion, or biases is not to develop even more technology, but to take a step back and ask "is this really beneficial in the first place, or are there other ways to conceptualize the issue at hand?" While some issues are best solved with policies, better stakeholder engagement, or changes in attitudes, some issues just cannot be solved easily.

(2) The *right to not be counted* is also important. For AI to take into account a whole range of parameters, the data can be used for sinister purposes. This is particularly relevant for cases where vulnerable people can find themselves listed in systems they do not consent to or want to be in. For example, databases counting sexual orientation can be dangerous in the hands of nation states with barbarian LGBTQ+ laws. For powerful governing forces looking to identify specific members of groups, the discontinuing and nondevelopment of systems rigged to produce results that disadvantage vulnerable groups is important.

(3) Do not treat AI as something magical without consequence, as Suchman and Suchman (2007) write, "We need to stop fetishishing tech. We need to audit algorithms, watch out for inequality, and reduce bias in computational systems, as well as in the tech industry". Technology like AI needs to be held accountable just like other powerful structural changemakers in society.

FURTHER RESOURCES

AI and diversity is a topic with growing scholarly and layperson interest. In this section, I provide a list of relevant places to look for further readings on the topic, including AI's connected field of robotics, which also impacts how we understand diversity questions for AI:

- The **AI for Good** is the United Nations' leading action-oriented, global, and inclusive platform on AI, created to accelerate the work toward meeting its Sustainable Development Goals.
- **AI4ALL**, a US-based nonprofit "dedicated to increasing diversity and inclusion in AI education, research, development, and policy [for opening] doors to artificial intelligence for historically excluded talent through education and mentorship." It is funded by Olga Russakovsky, Fei-Fei Li, and Rick Sommer from Stanford University.
- **AI in Africa**, a platform for forward-thinking initiatives focused on systematic change for disadvantaged communities in Africa.
- **AI-in-Asia.com**, a collection of perspectives on reframing AI governance in Asia, from the Singapore based Rule of Law Programme Asia and the German Konrad-Adenauer-Stiftung.
- The **AI Now Institute** is an interdisciplinary research center at New York University (NYU) in the US, aiming "to produce interdisciplinary research and public engagement to help ensure that AI systems are accountable to the communities and contexts in which they're applied." It was founded by Kate Crawford (previously at Microsoft) and Meredith Whittaker (previously at Google) in 2017.
- **AI4EU** aims to provide the latest news on cutting-edge AI applications, development trends, research, ethics, and social impact, by bringing together the AI community while promoting European values.
- The **Algorithmic Justice League** is a digital advocacy organization based in Cambridge, MA, USA. AJL's mission is to raise public

awareness about the impacts of AI, equip advocates with empirical research to bolster campaigns, build the voice and choice of the most impacted communities, and galvanize researchers, policymakers, and industry practitioners to mitigate AI bias and harm. It was founded by Joy Buolamwini in 2016.

- **ATONATON** is a research studio inventing better ways to live with machines, combing research lab innovation and designing studio ingenuity building prototypes of alternative futures. It is led by Madeline Gannon, in the Greater Pittsburgh Area, US.

- **DataKind** "focused on bringing data science [to make a] sustainable planet in which we all have access to our basic human needs," funded by Jake Porway, with a New York HQ.

- **Data Science for Social Good** aims to "create and sustain communities, programs, and solutions that enhance the use of responsible data science and AI for equitable social good" at the University of Chicago.

- **Data & Society** is an independent nonprofit research organization in New York, with the mission of producing "original research on topics including AI and automation, the impact of technology on labor and health, and online disinformation" and "convening "researchers, policymakers, technologists, journalists, entrepreneurs, artists, and lawyers to challenge the power and purpose of technology in society." It is funded by danah boyd (partner researcher at Microsoft and visiting professor at NYU).

- The **Data Privacy Lab** in the Institute for Quantitative Social Science at Harvard University "creates and uses technology to assess and solve societal, political and governance problems, and teaches others how to do the same." It is led by Latanya Sweeney.

- The **Foundation for Responsible Robotics** is a nonprofit, nongovernmental organization headquartered in The Hague, Netherlands with a mission "to shape a future of responsible robotics and artificial intelligence (AI) design, development, use, regulation, and implementation." It is led by Aimee van Wynsberghe and Noel Sharkey.

- **Robotics4EU** is a Coordination and Support project funded by the European Commission of Research, investigating the impact of robots on society "mapped from a social-humanities perspective, with a focus on how different societal stakeholders are impacted by robotics for their own lives and sectors." It is led by Anneli Roose at CIVITTA Estonia.

- The **Social and Intelligent Robotics Research Laboratory** at the University of Waterloo investigates human–robot interaction, cognitive and developmental robotics, and embodied artificial intelligence. It is led by Kerstin Dautenhahn and Chrystopher Nehaniv.

SUMMARY

This book has looked at how AI systems are connected to diversity among their users and producers and how diversity thinking can impact AI systems in return. I have discussed how AI practices are situated in sociotechnical networks and structures and are not detached from the world in which they are constructed. Awareness of who benefits from AI and who is excluded is crucial for moving toward a fairer use of AI that benefits more diverse groups of people. AI systems can be used as a driver for exclusion mechanisms, for surveillance and control of "others" and undesirables in society and systems alike. However, total fairness, obliteration of bias, and no exclusions are impossible. A computer is built to discriminate between two variables, 0 and 1, and from there, choices are made on what to prioritize. Some bias will always exist in a world where resources, time, and data are nonabundant, and priorities have to be set. This does not mean that we should accept any bias, and certainly not unfair decision-making, whether it be from humans or computer systems!

Through responsibility, awareness, and focused plans for diversity thinking, AI can be utilized as a changemaker for inclusion and a building tool for a better future. By breaking the black boxes and demystifying AI as something built by humans, for humans, we can

ensure that it doesn't exclude groups of people based on their individual attributes. Although AI is narrow, human access to it should be general, for everyone. My hope is that you, by reading this book, will work toward a more diverse and inclusive sociotechnical world of AI—starting with your own practices and organizations, one algorithm at a time.

REFERENCES

Akrich, M., Callon, M., Latour, B., & Monaghan, A. (2002). The key to success in innovation part I: the art of interessement. *International Journal of Innovation Management*, 6(02), 187–206.

Aldrin Salskov, S. (2020). A critique of our own? On intersectionality and "Epistemic Habits" in a study of racialization and homonationalism in a nordic context. *NORA-Nordic Journal of Feminist and Gender Research*, 28(3), 251–265.

Berker, T., Hartmann, M., & Punie, Y. (2005). *Domestication of media and technology*. McGraw-Hill Education.

Broussard, M. (2018). *Artificial unintelligence: How computers misunderstand the world*. MIT Press.

Buolamwini, J. A. (2017). *Gender shades: intersectional phenotypic and demographic evaluation of face datasets and gender classifiers*. MIT.

Crawford, K. (2021). *The atlas of AI*. Yale University Press.

Crenshaw, K. (1990). Mapping the margins: Intersectionality, identity politics, and violence against women of color. *Stanford Law Review*, 43, 1241.

Cuadraz, G. H., & Uttal, L. (1999). Intersectionality and in-depth interviews: Methodological strategies for analyzing race, class, and gender. *Race, Gender & Class*, 6, 156–186.

Darling, K. (2016). Extending legal protection to social robots: The effects of anthropomorphism, empathy, and violent behavior towards robotic objects. In Ryan Calo, A. Michael Froomkin, & Ian Kerr (Eds.), *Robot law* (pp. 213–232). Edward Elgar Publishing.

Dressel, P., Minkler, M., & Yen, I. (1997). Gender, race, class, and aging: advances and opportunities. *International Journal of Health Services*, 27(4), 579–600.

Fallan, K. (2008). De-scribing design: Appropriating script analysis to design history. *Design Issues*, 24(4), 61–75.

Gellers, J. C. (2020). *Rights for Robots: Artificial Intelligence, Animal and Environmental Law* (Edition 1). Routledge.

Gendered Innovations (2020). *Intersectional approaches*. Retrieved February 2, 2022 from http://genderedinnovations.stanford.edu/methods/intersect.html

Gunkel, D. J. (2018). *Robot rights*. MIT Press.

Haraway, D. (1988). Situated knowledges: The science question in feminism and the privilege of partial perspective. *Feminist Studies*, 14(3), 575–599.

Lovett, R., Lee, V., Kukutai, T., Cormack, D., Rainie, S. C., & Walker, J. (2019). Good data practices for Indigenous data sovereignty and governance. In Angela Daly, Monique Mann, & S. Kate Devitt (Eds.), *Good Data*, 26–36. https://books.google.no/books?hl=en&lr=&id=Y0vUDwAAQBAJ&oi=fnd&pg=PA8&dq=Good+data+(2019+devitt&ots=hraWipl8Jl&sig=lAr3kXxmsDbjKrb wuFi17Jev5_o&redir_esc=y#v=onepage&q=Good%20data%20(2019%20 devitt&f=false.

Mahoney, A. D., Cuellar, M. G., Johnson-Ahorlu, R. N., Jones, T.-A., Warnock, D. M., Clayton, K. A., Ray, V. E., Lee, E. M., Maynard, T., & Goerisch, D. (2019). *Intersectionality and higher education: Identity and inequality on college campuses.* Rutgers University Press.

Margulies, P. (1994). Asylum, intersectionality, and AIDS: Women with HIV as a persecuted social group. *Georgetown Immigration Law Journal*, 8, 521.

May, V. M. (2015). *Pursuing intersectionality, unsettling dominant imaginaries.* Routledge.

Oudshoorn, N., Rommes, E., & Stienstra, M. (2004). Configuring the user as everybody: Gender and design cultures in information and communication technologies. *Science, Technology, & Human Values*, 29(1), 30–63.

Perez, C. C. (2019). *Invisible women: Exposing data bias in a world designed for men.* Random House.

Rettberg, J. W. (2020). Situated data analysis: a new method for analysing encoded power relationships in social media platforms and apps. *Humanities and Social Sciences Communications*, 7(1), 1–13.

Runyan, A. S. (2018). *What is intersectionality and why is it important?* American Association of University Professors. Retrieved February 5, 2022 from https://www.aaup.org/article/what-intersectionality-and-why-it-important

Smith, J. K. (2021). *Robotic persons: Our future with social robots.* Westbow Press.

Smith, J. K. (2022). *Robot Theology: Old Questions through New Media.* Wipf and Stock Publishers.

Søraa, R. A., Nyvoll, P., Tøndel, G., Fosch-Villaronga, E., & Serrano, J. A. (2021). The social dimension of domesticating technology: Interactions between older adults, caregivers, and robots in the home. *Technological Forecasting and Social Change*, 167, 120678.

Sørensen, K. H. (1994). *Technology in use: Two essays in the domestication of artefacts* (Centre for Technology and Society Working Paper 2); p. 94). Centre for Technology and Society.

Suchman, L., & Suchman, L. A. (2007). *Human-machine reconfigurations: Plans and situated actions.* Cambridge University Press.

Vardeman, J., & Sebesta, A. (2020). The problem of intersectionality as an approach to digital activism: the Women's March on Washington's attempt to unite all women. *Journal of Public Relations Research*, 32(1-2), 7–29.

Verbeek, P.-P. (2005). Artifacts and attachment: A post-script philosophy of mediation. In Hans Harbers (Ed.), *Inside the politics of technology* (pp. 125–146). Amsterdam University Press.

Weber, L., & Fore, M. E. (2007). Race, ethnicity, and health: An intersectional approach. In *Handbooks of the sociology of racial and ethnic relations* (pp. 191–218). Springer.

INDEX

Printed in the United States
by Baker & Taylor Publisher Services

Printed in the United States
by Baker & Taylor Publisher Services